WRECK CHASING 2
Commercial Aircraft

**NICHOLAS A. VERONICO · ED DAVIES
DONALD B. MCCOMB JR. · MICHAEL B. MCCOMB**

This book is dedicated to the memory of those passengers and crews who made the ultimate sacrifice. Yet their loss was not in vain because of the vast improvements in aviation safety and air traffic control procedures resulting from these accidents. They should not be forgotten.

Published by

| WORLD TRANSPORT PRESS, INC. | P.O. Box 521238
Miami, Fla. 33153-1238
Tel. +1 (305) 477-7163
Fax +1 (305) 599-1995 |

Copyright 1996 by Nicholas A. Veronico, Ed Davies, Donald B. McComb, Michael B. McComb.
All rights reserved.
ISBN 0-9626730-3-X
First Edition July 1996
Printed and bound in the United States of America

No part of this publication may be reproduced, stored in a retrieval system, or transmitted by any means without first seeking written permission of the publisher.

The publisher acknowledges that certain terms, names, and model designations are property of the trademark holder. They are used here for correct identification purposes only.

All information contained in this volume is accurate at the time of publication.
No guarantee is made to the existence, location, or condition of any aircraft wreck.
This volume is intended to be a guide for research purposes only.

The title **WRECK CHASING** is copyright 1993 by Pacific Aero Press,
P.O. Box 20092, Castro Valley, Calif. 94546-8092, and is used here with permission.

ON THE FRONT COVER
JERRY BOAL INSPECTS THE TAIL SECTION OF N1240N ON PANTHER PEAK IN THE SEQUOIA NATIONAL PARK.
(G. PAT MACHA COLLECTION)

ON THE BACK COVER
CLOCKWISE FROM TOP RIGHT: THE TAIL SECTION OF TWA DC-3 N1946 RESTS INVERTED ON THE SLOPES OF MT. POTOSI. THIS AIRCRAFT CARRIED ACTRESS CAROLE LOMBARD AND 21 OTHERS TO THEIR DEATHS ON JAN. 16, 1942. (UPI/BETTMAN) RPM GAUGE FOUND AT THE CRASH SITE; ONE OF THE DC-3'S ENGINES RESTS IN THE CLEAR; ARTIFICIAL HORIZON RING FROM N1946. (PHOTOS BY MICHAEL B. MCCOMB)

World Transport Press offers a full line of aviation related books, magazines, calendars, and much more. For a free, four-color catalog, please send your name and address to:
World Transport Press
P.O. Box 521238, Miami, Fla. 33153-1238
Tel. +1(305) 477-7163 • Fax +1 (305) 599-1995

CONTENTS

Acknowledgements
Preface

Part I

Wreck Chasing: Locating Commercial Aircraft Crash Sites.. 6
  By Nicholas A. Veronico
Researching An Airliner Crash From Start To Finish.. 9
  By Nicholas A. Veronico

Part II

Overdue From Kingman: The Crash of Fokker F-10A NC279E.. 14
  By Nicholas A. Veronico
Restricted Medical Certificate Ignored: Robin Air Lines C-46.. 17
  By Nicholas A. Veronico
Fuel Transfet Starts In-Flight Fire: The Crash of United DC-6 N37503... 23
  By Michael B. McComb
Collision Course: New York Mid-Air.. 28
  By Donald B. McComb Jr.
The DC-3 That Crashed And Flew Again... 34
  By Ed Davies
Final Destination: Grand Canyon Mid-Air... 39
  By Donald B. McComb Jr. and Michael B. McComb
Cross Country To Disaster: All On Board United Flight 615 Perish.. 47
  By Nicholas A. Veronico
Severe Turbulence Ends BOAC Flight 911 Over Mt. Fuji... 52
  By Donald B. McComb Jr.
Lightning Strike Ignites Explosion — Downing Pan Am Flight 214.. 55
  By Donald B. McComb Jr.
The Mystery Of TWA Flight Three: Actress Carole Lombard Perishes In Crash............................ 64
  By Donald B. McComb Jr. and Michael McComb
Ill-Fated Airliners: Photo Gallery... 70
Training Flight Mid-Air: Cal Eastern and Overseas National... 80
  By Nicholas A. Veronico
Christmas Eve Disaster: Flying Tigers Constellation.. 84
  By Nicholas A. Veronico and Ed Davies
Crash Shorts: A Trio of C-46 Crashes.. 88
  By Nicholas A. Veronico

Part III

Appendix I—   Jet And Turbo-Prop Crashes... 94
Appendix II   —   Propliner Crashes... 109
Appendix III   —   DC-3 Crashes In The United States.. 120
Appendix IV   —   Obtaining A USGS Topo Map... 124
Appendix V   —   Further Sources... 124
Appendix VI   —   Work Sheet... 125

Bibliography And Suggested Reading... 126
About The Authors... 127

ACKNOWLEDGEMENTS

This volume would not be possible without the help of dozens of people who have generously shared their time, experiences, and collections with the authors. Special thanks to William T. Larkins and G. Patric Macha for providing essential photographic and research materials. Thanks also to: *Airliners* magazine Assistant Editor Jon Proctor; Vera Lynn Davids; Heidi Van Zant; Ron Strong; Scott Thompson; Ian Abbott; Ken Miller; Tom Northrop, Pacific Aerial Surveys; William and Sally Dell; Violet Andrada; Arue Szura; Betty Anderson; Earl Holmquist; Bob Hofford; Christopher Schiess; John T. Wible, Aireports; Todd Hackbarth; Taigh Ramey; Bill Hough; Linda Zupcic; Dr. Lee Schaller; Don Gridley; Arthur Pearcy; Eleanor Rose; J.R. Lyons; Bob Swanson; Milo Peltzer; Gary Fong, *San Francisco Chronicle*; Jim Rowan; Stephen L. Griffin; Brian Baum, Museum of Flight; Clint Groves, ATP/Airliners America; John M. Campbell; A. Kevin Grantham; Brian Sullivan; Matt Heller, Wildflower Productions; Wayne McPherson Gomes; Tony Veronico; Armand Veronico; L. Ray; Caroline Bingham; Karen Haack; Robert A. Kropp; Roy D. Meyers; Keith and Pat Armes; Bruce and Pauline Drum; Oscar Bidopia; Rosy Medina; Sergio Bressen; Kris Martinsen; John Rodriguez; Ramano Urbat; and John Harjo.

Nicholas A. Veronico
Ed Davies

I would like to thank the following: the staff at the NTSB and the Accident Investigaiton Branch, Farnborough, England; Mr. R.A.R. Wilson of the Historical Aviation Service; retired 707 captains Peter J. Howard, Peter J. McKeown, and Eric Pritchard; Robert Serling, whose books on air safety and the airline industry set the standard for well written aviation books; fellow editor and friend Jon Proctor, whose vast store of commercial aviation knowledge provided details of the 1956 TWA/UAL accident; and for years of putting up with the odd bits of airframe I sometimes leave lying about the house, thanks also go to friends Jeff S. Johnson and Dr. Dana X. Kerola for tolerating my esoteric collection.

Donald B. McComb Jr.

A note of appreciation for their assistance to: the NTSB; National Technical Information Service; General Microfilm; the FAA; Bryce Canyon National Park; USGS; Bettman Archives; Donald Simonis and the Bureau of Land Management; George R. Perkins and the U.S. Forest Service; retired TWA captain Arthur E. Kelly; Dick Wallin; Jon Proctor; Craig Fuller; Robb Hill; Doug Scroggins; Scott Featherstone; William Walldock and Embry Riddle Aeronautical University; Harvey Gardner of Page Steel; James Thompson; Robert J. Serling; G. Pat Macha; Terry Geary of Wreckfinders; and Robert Brackett.

Michael B. McComb

PREFACE

Airliners are the most majestic birds in the sky today. When one crashes, it is a tragedy. Despite the loss of life, many early airliner accidents led to positive changes in our air traffic control system and improved airliner designs.

Locating a crash site can be a challenging and moving experience. The sight of a former airliner broken into hundreds of small pieces can be overwhelming. Crash sites are extremely fascinating and one can spend hours exploring and mapping a location.

This volume takes the reader through the necessary steps required to locate the final resting places of numerous commercial aircraft. A worksheet has been provided as a starting point for research organization. A number of appendices will provide additional tools necessary to begin or complete a Wreck Chasing adventure. Finding an aircraft crash site is truly 90 percent research and 10 percent hiking.

Crashes profiled range from controlled flight into terrain, the failure to see and avoid that has resulted in mid-air collisions, to airliners that have been forced out of the sky by Mother Nature. All airliner crashes are not catastrophes as evidenced by The DC-3 That Crashed And Flew Again and Appendix III DC-3 Crashes In The United States. It is hoped that this listing of DC-3 incidents will save Wreck Chasers from working long and hard at researching a crash, only to be disappointed at the outcome. The listing of piston, turbo-prop and jet crashes has been narrowed down to sites where wreckage may still remain. Also included are aircraft that have crashed into bodies of water.

Follow the steps presented, ask questions, and complete your research. Use common sense and be prepared for the hike. An interesting adventure in an often overlooked facet of aviation awaits.

Wreck Chasing: Locating Commercial Aircraft Crash Sites

Decades after an airliner has crashed, its final resting place is an interesting sight. What was once tons of twisted metal now lies scattered over a wide area. Some parts are recognizable, others are buried under the soil. All were once part of a flying airliner.

Locating a commercial aircraft crash site requires more research than any other type of aircraft crash. Many times the entire wreck site has been swept clean, whether by a government safety organization, an insurance company, or the land owner. Thus, finding pieces of an actual aircraft are much like drilling for oil--sometimes you come up with a dry well. The fun is in locating the exact site, documenting your search, and hopefully finding an airliner's final resting place.

Choosing Your Site

In Volume I, *Wreck Chasing: A Guide To Finding Aircraft Crash Sites,* readers visited the locations of the British Commonwealth Pacific Airlines DC-6 VH-BPE, msn 43125, and Pan Am's Martin M-130 "Philippine Clipper." Both crash sites held substantial amounts of wreckage, partly due to their inaccessibility and the inhospitable terrain. The sites were chosen first on their geographic location: the BCPA DC-6 was a day trip, while the "Philippine Clipper" was an overnighter. The second consideration was also of a geographic nature. Researching the crash would require visits to local libraries, museums, and to conduct interviews with the participants. For financial considerations these destinations needed to be close to home. The third consideration was the aircraft's type. The M-130 was a rare bird while the BCPA DC-6 was the ultimate in early 1950's air travel. Both represented types that are now extinct in passenger carrying service — the M-130 crash being the only known surviving parts of the type.

When looking for a crash site to explore, first try to choose an aircraft type you are interested in. If you have no interest in a particular type, then why go? Researching a crash takes time and patience, so it helps if the aircraft type holds your interest. Review Appendices I, II, and III for a listing of commercial aircraft that have crashed. Use this as a starting point for your search.

Beginning Your Search

Ninety-nine percent of the effort required to find a crash site is expended doing research. Careful planning will yield positive results and spare you from a pointless hike.

The following outline provides the recommended search path for locating a commercial aircraft crash site.

Step I	Decide on the type of aircraft to search for.
Step II	Prepare Worksheet. See Appendix VI.
Step III	Obtain Documentation. Begin with the aircraft's accident report.
Step IV	Obtain newspaper or TV reports of the crash.
Step V	Obtain a topographical map of the crash site.
Step VI	Interview the participants.
Step VII	Locate the crash Site. Plan your trip.
Step VIII	The Crash Site.
Step IX	Making your research available to others.

Taking The First Step

Upon deciding on the specific aircraft crash site to investigate, prepare a worksheet. Fill in the make/model, registration or serial number at the time of the crash, any previous registrations, the manufacturer's serial number, and the plane's date of delivery to its first operator. Check whether it was a civilian airliner and note the airline name, or if military, fill in the branch of service and unit assigned. Some of this information is available in Appendices I, II, and III, or can be located in a number of reference books (See Bibliography/Suggested Reading). Next note the location of the wreck and if you have a more specific idea of where the plane when down, list the 7.5 minute quadrangle map on which the site is located (See Appendix IV, How To Obtain A Topo Map).

Your local community or college library is your next destination. Commercial aircraft crashes receive huge amounts of attention from the press. Nothing sells papers or hooks TV viewers into tuning in the late news like a screaming crash headline or sound bite. Start with the microfilm section and begin with the newspaper published closest to the crash site. Your search should begin with the edition following the crash. On the worksheet note the name of the newspaper and the pages you find related articles. Copies of these pages

PART I

should be added to your file. Take time to look through each day's newspaper for next two weeks to one month. See if there is any mention of wreckage removal. Depending on your library, also check the *Los Angeles Times, New York Times,* or the major market daily in your area. These papers usually cover all aviation disasters.

The newspaper will also provide you with a location of the crash, the names of emergency personnel who responded to the scene, and the names of local residents who witnessed the crash. A check of the local phone book will yield interesting results. Many of the names mentioned in the newspaper will still be residents of the local or surrounding area, sometimes more the 40 years after a crash. Note these names on the worksheet. Show their current address if you are able to locate anyone. Fill in the location or the directions to the crash site given by the newspaper. Depending on the area and vintage of the crash, the newspaper may report "a road had to be cut to the crash site in XYZ canyon." This is important information that can be transferred to a topo map to help pin point the location.

Obtaining Crash Reports For Commercial And Military Transports

Send for, or obtain commercially (See Appendix V, Additional Sources Of Information), the official aircraft crash report. The date of the crash, location, aircraft type, and registration must be in your request for information. The National Transportation Safety Board may charge for copies of crash reports. If the plane was operated by a commercial airline, write to:

National Transportation Safety Board
800 Independence Ave., S.W.
Washington, D.C. 20594

For Air Force aircraft, the crash reports are held at the Air Force Inspection and Safety Center. Your request will need the crash date, location, aircraft type, and serial number if known. Write to:

Headquarters, Air Force Inspection & Safety Center
Kirtland Air Force Base, N.M. 87117

The Air Force censors the crash investigator's opinions from the reports to keep their opinions private. The censor's marks do not interfere with the location information, but be prepared to see a lot of black ink.

The Navy keeps reports for crashed aircraft at the Washington Navy Yard. Individual crash reports may be obtained, usually at not cost. They also offer reports for older aircraft types on microfilm (See Volume I, *Wreck Chasing: A Guide To Finding Aircraft Crash Sites* and Appendix II, *U.S. Navy Pilot History Cards of Aircraft Accidents,* Page 69). To order individual aircraft crash reports, U.S. Navy aircraft history cards, or crash reports on microfilm, write to:

Department of the Navy
Naval Aviation History and Archives
Building 157-1
Washington Navy Yard
Washington, D.C. 20374-5059

Douglas C-54G-10-DO Transocean Air Lines N88942

Two very different views of Transocean Air Lines' C-54G-10-DO N88942. Above: A sistership to N88942, N90915, taxies at Oakland, California, Transocean's home base. N88942 was en route from Walker Air Force Base, Roswell, New Mexico, to Guam with a single stop at Oakland. Thirty maintenance personnel from the 509th Bomb Wing and a crew of five, including Transocean's chief pilot Harvey W. Rogers, perished in the March 20, 1953 crash. (William T. Larkins) Below: N88942 as no airplane should look. Transocean mechanic Wm. "Bill" Dell took this photo shortly after the crash. The field was part of Andrada's Dairy near the intersection of Whipple and Alguire Roads, in Alvarado, California.

Researching An Airliner Crash From Start To Finish

For this example we will research the crash of Transocean Air Lines C-54G-10-DO N88942, msn 36076. The C-54 was the military version of the DC-4 airliner and was 93 feet 10 inches long, with a wing span of 117 feet 6 inches. It had a maximum speed of 280 knots at 23,000 feet. C-54G-10-DO is the 'G' model C-54, modification block 10, constructed by Douglas at Santa Monica, California.

Starting with the worksheet, fill in the above information; make model C-54G-10-DO; Registration/serial N88942; Operator Transocean Air Lines; crash date March 20, 1953.

If you desired, the civil history of this aircraft could be obtained by sending the aircraft type, registration number, and msn to:

Federal Aviation Administration
P.O. Box 25504
Oklahoma City, OK 73125-0504

If these records can be located in storage, you will receive a microfiche copy of the history of N88942. Another route is to look up the aircraft history in books such as A.B. "Tony" Eastwood and John Roach's Piston Engine Airliner Production List (See Bibliography/Suggested Reading). Presented by aircraft type in msn order, this is book provides a summary of the FAA information — without the wait. Piston Engine Airliner Production List shows that N88942 was delivered to the U.S. Army Air Forces on October 9, 1945 as serial number 45-0623. The plane was then leased by the Army Air Force to Pan American World Airways sometime in 1946. (The FAA records would provide the exact date, the books are sometimes vague on detailed points like this.) Pan Am christened the plane Clipper Bostonian. In 1951, the Air Force leased the plane to Transocean Air Lines. The carrier used the plane exclusively for Military Air Transport Service (MATS) contract flights. Within two years the plane had crashed on approach to the Oakland Airport on March 20, 1953, in the town of Alvarado, California.

The town of Alvarado is located on the Newark, California quadrangle. The Civil Aeronautics Board (CAB) adopted its finding of the crash on October 9, 1953, and published the report on October 14, 1953. The report number is 1-0016. Enter the above information on the worksheet.

Now that the top half of the worksheet has been completed, the next step is to review the newspaper. Three major papers were published within a 50 mile radius of the crash site. They are the *San Francisco Chronicle*, Oakland *Tribune*, and the *San Jose Mercury News*.

Headlines read: "Wing Ice May Have Caused DC-4 Crash That Killed 35," *San Francisco Chronicle*; 35 Die In Eastbay Plane Disaster, Oakland-Bound DC-4 Jammed With Air Force Men, Falls and Burns In Barley Field Near Alvarado," Oakland *Tribune*. The *San Francisco Chronicle* coverage was good, but for a Wreck Chaser, the information in the Oakland *Tribune* was a gold mine.

The Flight, The Crash, And The Newspaper Accounts

Operating as Transocean Air Lines' Flight 942, N88942 left Walker Air Force Base, Roswell, New Mexico at 12:11 p.m. MST, en route to Oakland, with a final destination of Guam. All 30 passengers were maintenance personnel from the 509th Bomb Wing — noted for dropping the atomic bombs on Japan. The flight was crewed by five. They were Harvey W. Rogers, 41, pilot, was Transocean's chief pilot; Herman E. Hum, 41, also a Transocean captain served as relief pilot; Fred W. Patchett, 37, copilot; Velma Sandridge, 27, and Lucile M. Chapman, no age available, flew as stewardesses.

Transocean employees examine the wreckage of N88942 in an Alvarado Barley Field. William "Bill" Dell croutches down.
(William Dell)

The flight reported passing over Winslow, Arizona, at 2:51 p.m. MST, where it refiled for an Instrument Flight Rules (IFR) Flight Plan. At 5:32 p.m. PST, Transocean Flight 942 checked in over Fresno, California, flying at 8,000 feet. Shortly thereafter, the flight was cleared into the Bay Area, first to report over

N88942's crash site then and now. Above: Following the crash, Transocean sent a number of workers to help with the investigation and clean-up. From left to right: Transocean employees William "Bill" Dell, Wendell Davies, Mike Lewis, and unidentified. (William Dell Collection) Below: The crash site is a still a ploughed field to this day. The lot is currently up for sale. Arrows indicate geographic feature used as a reference point when locating the crash site. Compare the two photos and notice how little has really changed in 40 years. (Nicholas A. Veronico)

Evergreen (near San Jose), then to contact Oakland Approach Control at the Newark radio beacon (fan marker and compass locator). Arriving at 6:09 p.m., over Evergreen, the flight was cleared to Newark, remaining at 8,000 feet. Less than a minute later, Flight 942 requested 7,000 feet but was denied due to traffic.

It was a drab spring day inside the bay. The CAB report states that the weather was "Measured ceiling 1,300 feet broken, 1,800 feet overcast, light rain, fog, visibility two and one-half miles, wind south-southwest 17 miles per hour, altimeter setting 30.09." Not real bad flying weather, not the best. Nothing that should have given a large four-engine airliner any problems.

At 6:19 the flight reported over the Newark radio beacon and was instructed to hold. Eight minutes passed before Oakland Approach Control cleared the flight to descend in the holding pattern to cross the Newark radio beacon at 3,500 feet, reporting at each 1,000 foot level, and then make a straight-in approach to land. The flight reported beginning its descent from 8,000 feet at 6:30 and six minutes later stated that it had crossed the Newark radio beacon inbound for landing. That was the last transmission from the flight.

"At approximately 6:38, the aircraft crashed in a barley field. Impact and fire destroyed the aircraft. There were no survivors," the CAB report states. "Investigation revealed that the aircraft crashed in a large flat field located three miles on a magnetic bearing of 323 degrees from the Newark compass locator and one and one-half miles northeast of the town of Alvarado, California. The surrounding terrain consisted of flat farm land on which were a few scattered houses, fences, and trees. The elevation of the field is approximately 17 feet MSL.

"The aircraft first struck the ground on its right wing tip and with the wing in a near vertical position, then cartwheeled and disintegrated. Wreckage was scattered over an area approximately 800 feet long and 300 wide. Due to impact forces and the resultant fire, the aircraft broke into numerous small pieces, many of which were destroyed by fire with only two large sections of the aircraft remaining after the crash. These were the fuselage center section with a portion of the left wing attached which was found lying inverted approximately 634 feet from the point of first impact and the rear section of the fuselage, including empennage, from approximately Station 660 rearward to the tail cone, which was also lying nearby."

On Page 5 of the March 21, 1953, edition of the Oakland *Tribune*, witness William Silva, a resident of Whipple Road, reported finding "pieces of ice up to eight inches long and three inches thick with smooth rounded inner sides." This ice was presumed to have come from the leading edge of the C-54's wing. (Although the cause of the crash has never been determined, icing was suspected as a major factor. The CAB report said, "One pilot, who was holding over Newark at 8,000 feet approximately 35 minutes after the accident occurred, reported encountering severe icing conditions and mild turbulence with ice approximately three inches in diameter accumulating on antenna masts. He said that the ice began to melt when the 4,500-foot level was reached in the descent.") The paper also reported that Silva's house was southwest of the crash site. Silva and another witness Henry

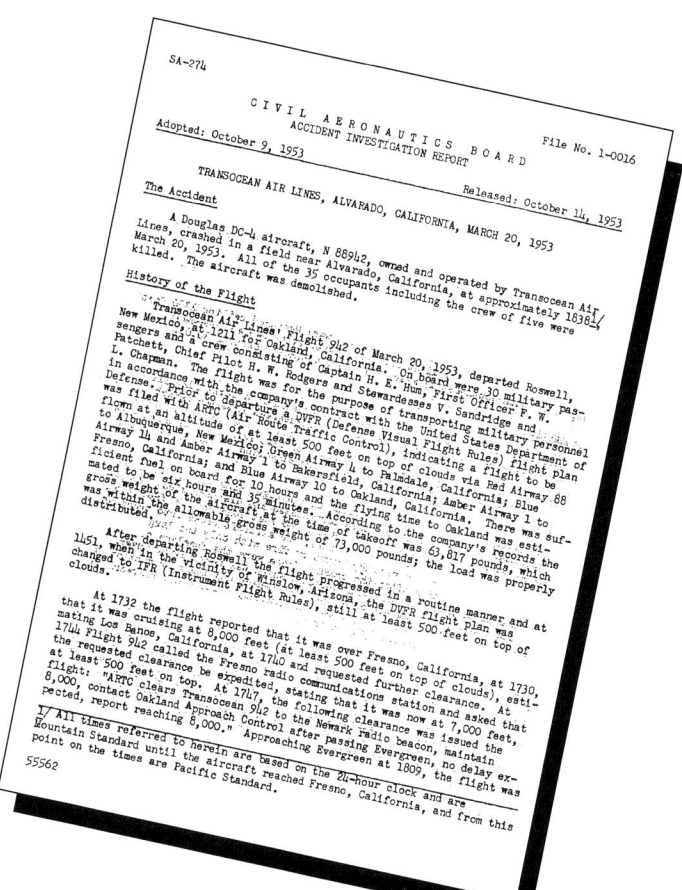

The Civil Aeronautics Board issued its final report on the crash on October 14, 1953. On the day of the crash, the plane was being crewed by a very experience crew. The plane was low time, having flown only 5,976 total hours. Although undetermined, the report lists one possibility for the crash as "the accumulation of ice on the surfaces of the aircraft in sufficient magnitude to have caused loss of control..." (Arue Szura Collection)

Gomes gave first-hand eye-witness accounts of the crash to the newspaper. Each wrote a column. Other witnesses named in the paper were Robert Lemos and Mrs. Edith Whipple. William Silva's wife, Wilma, was also listed.

The *Tribune* gave the location of the crash as "a fourth of a mile northeast of the intersection of Alguire Road and Whipple Road on the ranch operated by Frank and Rick Andrada."

Adding To The Worksheet, Interviewing The Participants

From the information in the CAB report and the newspapers, names of the witnesses and a location of the wreck have been learned. After entering this information on the worksheet, start with the local phone book. Gomes, Lemos, and Whipple were not listed. Calls to others with the same last names could yield family members or next of kin, but that step was skipped after the Silva family was located. They had moved from their home near the crash site, but continued to work and raise a family in the community. Silva pro-

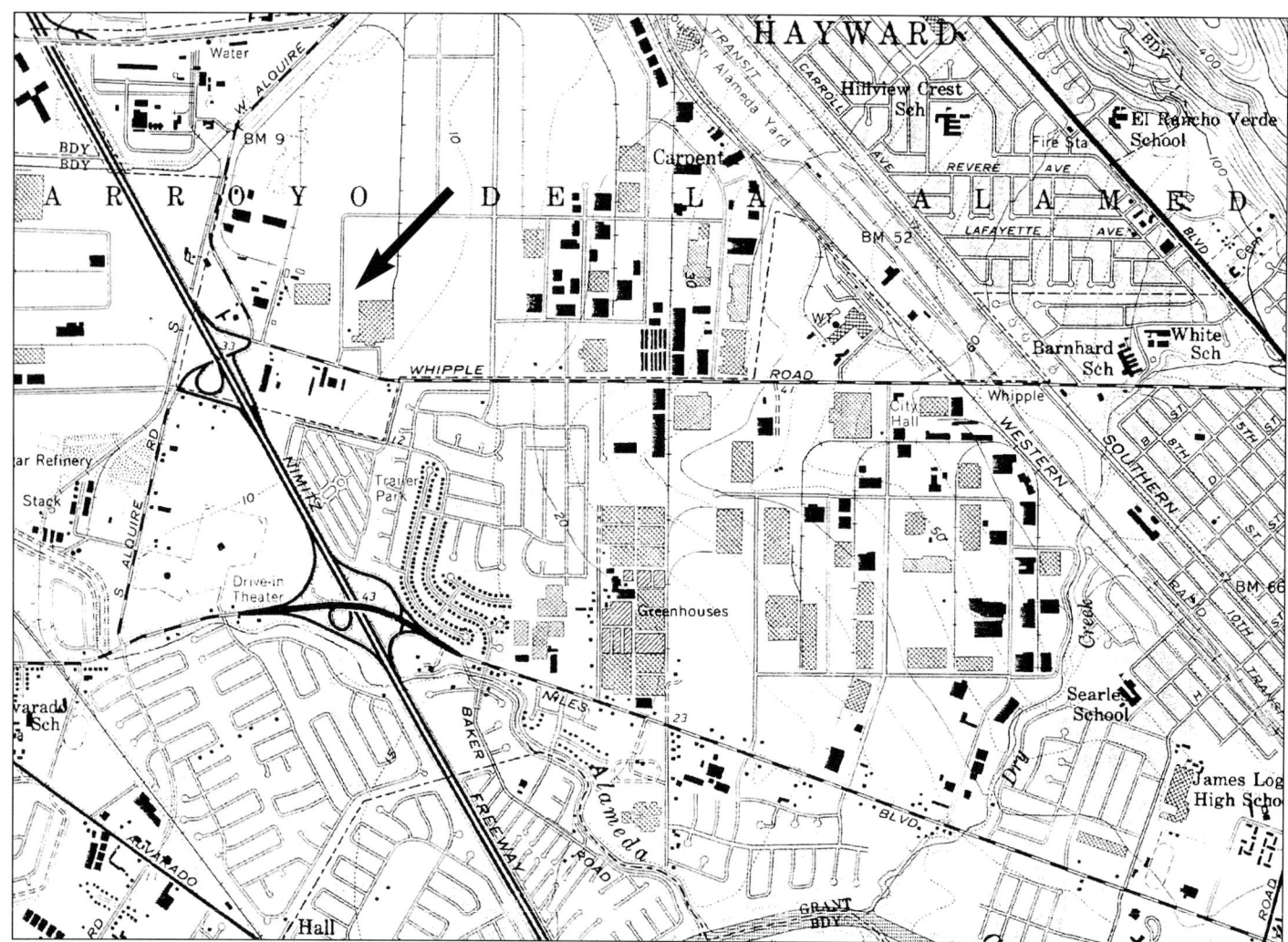

U.S.G.S. topo map showing Hayward and the Alvarado and Decoto districts. N88942 crashed off Whipple Road near Highway 880. The crash location is highlighted by an arrow. (U.S. Geological Survey)

vided more clues to where the wreck site was, contrasting the area in 1953 — before Interstate 880 was built, to the area today. "We lived where Walmart is now and the accident happened across the freeway, maybe a quarter of a mile away. The crash was on the east side of the freeway. We were in the house just before dinner. We heard this roar, the plane passing overhead. I looked outside and saw the plane going over our house and it was tilted to the left. The left wing hit the ground and then it burst into flame. I ran across the street [Whipple Road] to this field and tried to find somebody alive, to no avail. The plane disintegrated when it hit. When I came back I found some pieces of ice, like they came off the wing, which I reported at the inquest," Silva said.

The location shown in the newspaper stated that the plane crashed on a ranch run by Frank and Rick Andrada, known as Andrada's Dairy. A review of the local phone book showed a listing for H. Rick Andrada. Unfortunately, he passed away in 1987. He is survived by his wife Violet Andrada. She said, "The ranch was on the corner of Whipple and Alguire. It covered a large area and had a dairy. At the time of the crash, they had just plowed and planted barley. It had been raining something terrible. My brother-in-law and Henry [Rick] took care of the ranch." She mentioned that her husband and brother-in-law leased the ranch from a gentle-

man named Wiegman who lived in Alvarado.

Mrs. Andrada said that after the crash, "Billy Silva [William, mentioned above] grabbed my hand and we jumped out there in the field to see who we could save. The wreckage was on the property, I imagine, about four months. Some of the plane's motors were in the ground six feet deep. The military dug them out and removed everything. They had a regular guard walking back and forth with guns over their shoulders. Anyone who was coming into the ranch to visit us was under surveillance. They had to show identification before the soldiers would let them in. The Army watched every little thing, they didn't miss a trick.

"Henry wrote a letter to every mother who wrote a letter to him. He told them that they had a minister, Catholic, Protestant, Jewish, right there at the site giving them the last rites. A lot of the parents wrote and Henry kept those letters for years and years."

Finding The Crash Site

Armed with this information, a street map was consulted to find the corner of Alguire and Whipple. How hard could it be to located a point one-quarter mile northeast of the intersection of Alguire and Whipple? Whipple is a major

Today the area around the crash site has been built up as an industrial park. The crash location is to the rear of this lot. Wiegman Ave. is named after the man who owned the property in the 1950s. (Nicholas A. Veronico)

street, it even has an offramp from Highway 880. The map showed an Alguire Parkway, but it was nowhere near and did not intersect Whipple Road. Now what?

Some original photos of the crash site would be helpful. Someone who was there and took photos would be even better. Author Arue Szura, who wrote *Folded Wings: A History of Transocean Air Lines*, was asked to locate someone who participated in this dark chapter of the carrier's history. Szura pointed the search towards William F. "Bill" Dell. Dell was a Transocean mechanic at the time and was assigned to help investigate the crash and comb through the wreckage. Dell's insights and photographs helped locate the crash site.

While plotting the gathered information onto the topo map, it was noticed that what is today Dyer Road was formerly S. Alguire Rd., and today's Industrial Blvd. was once W. Alguire Rd. Due to the cost of topo maps, consider photocoping or placing a sheet of tracing paper over the map before plotting any information. That will keep your map fresh and you can mark all over a copy if you need to. Using all of the information garnered from the newspapers, CAB reports, and the eyewitness accounts, a likely crash location was plotted. Upon arrival at the site, the photographs received from Dell were matched to the surrounding terrain. We had found it.

Unfortunately, there was no wreckage left. It had been an interesting search. If you would like to visit this location, take Highway 880 and exit at Whipple Rd. Drive east one-quarter mile to Wiegman (the street was named after the landowner). Turn left, pass the building on the right and stop at the vacant lot on the right. Walk back about 100 yards and look towards the hills. Match the photographs on these pages to the hills behind you. You are standing at the crash site.

The land is currently up for sale. Hopefully, once new owners have settled in, a plaque to honor the memories of those who perished can be placed at, or near the site.

Start your own Wreck Chasing adventure by follow the steps outlined here and in Volume I, *Wreck Chasing: A Guide To Finding Aircraft Crash Sites*. You will have an adventure to remember.

WC

Left: *Completed worksheet for Transocean C-54G N88942. This worksheet helps the Wreck Chaser gather information in an orderly manner.* (Nicholas A. Veronico)

Overdue From Kingman
The Crash of Western Air Express Fokker F-10A NC279E

By Nicholas A. Veronico

The Fokker F-10A differed slightly from the F-10 model shown above. Fokker F-10A NC279E was lost on its delivery flight while en route to Alhambra, California. The early airliner rests in a remote canyon near Southern California's Lake Arrowhead.
(Western Air Lines via Veronico Collection)

A new standard of luxury in air travel was introduced in April 1928 when Western Air Express began service with the 10-passenger Fokker F-10 trimotor. The air carrier was granted $180,000 in 1927 by the Daniel Guggenheim Fund for the Promotion of Aeronautics to promote commercial air travel. Three Fokker F-10s were purchased for use on "The Model Airway," a route between Los Angeles and San Francisco/Oakland. Eventually Western Air Express would operate a fleet of five F-10s and 20 F-10As.

Powered by three 450-horsepower Pratt & Whitney Wasp radial engines, the F-10A could carry 12 passengers over a range of 765 miles at a cruising speed of 123 mph. Huge by the standards of the day, the F-10A had a wing span of 79 feet 2 inches, a fuselage length of 50 feet 7 inches, and weighed in at 13,100 pounds. For passengers, in-flight "meals," operable windows for fresh air, and a rear cabin restroom increased the comfort level while airborne. For the flight crew, the cockpit boasted an instrument panel with lights for night flights and brakes for the main landing gear. Fokker eventually built 65 F-10s and 59 of the slightly larger F-10As. Western Air Express flew the F-10A until December 1935 when the last four of the type were sold from the fleet.

Overdue From Kingman

Fokker F-10A, NC279E, msn 1011, was delivered to Western Air Express on Feb. 2, 1929. It met an untimely end just over one year later on Feb. 23, 1930. While flying from Kingman, Arizona, to Alhambra, California, NC279E crashed eight miles northeast of Lake Arrowhead, California, during a severe snowstorm. Time of the crash was fixed at about 2:30 p.m. PST. The time was arrived at by checking the wrist watches of the crew—one stopped at 2:25 and a second at 2:27. The clock in the instrument panel continued to tick time off until stopping at 2:45. The cabin of the airliner was void of passengers and only the crew of three, including Western Air Express pilot James E. Doles, co-pilot A.W. Bieher, and steward John Slaton, were killed in the crash.

Investigators at the time surmised that pilot Doles flew from Kingman to the eastern side of the San Bernardino

Pat Macha lends scale to the Fokker F-10A's remains. The F-10A was powered by three 450-horsepower Pratt & Whitney Wasp engines and could cruise at 123 mph for 765 miles. Twelve passengers were accomodated within the airliner's 50-foot 7-inch fuselage. If this fuselage was recovered, wings would have to be rebuilt from plans. (G. Pat Macha)

Mountains where he encountered a snowstorm. While threading his way across the mountain range in the storm, Doles made the decision to turn back in an attempt to retrace his path to Kingman. He never made it.

From the damaged tree tops in the area, investigators guessed that Doles must have been blown down into the tree tops by a fierce gust of wind that weakened the wing's structure. Shortly after contact with the trees, the wing separated from the fuselage, fluttering to the ground. This sent the fuselage in a 45-degree dive through the trees, where it impacted the ground. The location of the fuselage's resting point was described as "an inaccessible canyon about eight miles northeast of Lake Arrowhead...halfway between Hokum Creek and Luna Mountain...and one mile from Mill Creek."

Fokker empennage in the brush. (G. Pat Macha)

Wing Located; Where's The Fuselage?

After the snowstorm cleared, the first sign of the wreckage was seen six days after the crash by Mrs. Juanita Eloise Burns. Burns had spent five days flying over the area in her 90-horsepower Travelair biplane looking for the wreckage. On the day the wreckage was discovered, she was riding as an observer with Dudley Steele, a pilot for Richfield Oil Company. Burns spotted the wreckage in a canyon, and the pair returned to report their findings.

The next morning at 2 a.m., a search party was formed to hike into the site. Burns and Steele led the way accompanied by Western Air Express' Fred Goodcell; and Ernest Shay and Ed Mitchell, both of whom were deputy sheriffs. The search party tracked over pine and spruce tree-studded ridges through a maze of canyons until dawn. As the sun rose over the tops in the east, Western Air Express pilot Monte Shelton circled over the wreckage to guide the searchers to the wing.

Upon locating the wing, they found it to be intact except for a few torn and missing pieces of fabric and a number of small holes in the leading edge. The wing tip was unscathed, putting to rest theories at the time that the plane's wing had struck a canyon wall, cartwheeling it into the ground. Although the wing had been located, no sign of the fuselage or the plane's three engines could be seen nearby.

Searchers fanned out in all directions, and two hours later volunteer searchers Lester Bugg, a Lake Arrowhead stone mason, and John Dexter, owner of the Squirrel Inn, signaled that they had discovered the engines, fuselage, and the remaining wing section. The fuselage had accordioned upon impact. The three engines were torn from their mounts and turned the wing into a twisted, broken mess. The bodies of the three crew men were packed out and then flown to Southern California's Alhambra Airport. From there the bodies were sent to their respective families for burial. After the crash investigation, the wreck was left on the mountain in what was considered a remote area in 1923.

The Crash Site Today

Due to the fabric-covered steel tube construction of the Fokker F-10A, there is no salvage value to the wreck site. However, the historical value is significant. Wreck authority G. Pat Macha is currently seeking a museum or interested parties who would like to recover the aircraft and begin the restoration process. Macha said, "The site is drivable with another hour or so walk up the mountain to reach the site. From the main highway, it is a long drive on dirt roads to get there. That's the primary hassle. The wreck is approachable only from the desert side of the mountain."

The F-10A sat relatively untouched until the 1950s when a fire went through the area. This damaged the site further, but the plane is considered worth recovering. Since the fire, the growth one would expect has not come back. Low scrub brush has returned, and although you must obtain permission, the trek to the site is considered worthwhile. Macha visited the site in the fall when the days were becoming short. He described the site, saying, "There is a big piece of fuselage with the vertical and horizontal tails still attached. The landing gears are there. The F-10A, besides having a tail wheel, had little skids. A skid was there. The wing had a heavy tubular metal frame with a wooden spar. There are wing ribs scattered about. The props are gone, but there are tons of engine parts and cowlings. We did not find any personal effects.

Harv LeBlanc with a wheel, cylinder, and other parts from Fokker F-10A NC279E. (G. Pat Macha)

"That F-10 was a big bird, so it's pretty impressive from that standpoint. The wreck is interesting from the early technology standpoint, that type of construction. We have one rib section and one wing strut, a cylinder head, and a three-blade prop hub—all of which we would donate to any museum who recovers the aircraft.

"If somebody wanted to make an effort to recover the aircraft, it would not be that tough helicopter-wise."

The restoration and recovery of this early example of our commercial aviation heritage needs to be undertaken by an experienced, well-financed organization. **WC**

Pat Macha with the twisted remains of the Fokker. The remains of one engine are in the center right of the photo. This aircraft could be recovered and restored by a museum with enough funds. (G. Pat Macha)

Restricted Medical Certificate Ignored
The Crash Of Robin Air Lines C-46F N8404C

By Nicholas A. Veronico

Robin Airlines Flight 416W departed New York on April 17, 1952, at 9:09 p.m. EST on a cross-country trip to the Los Angeles suburb of Burbank. The plane never made its final destination. Trip 416W ended in the La Puente Hills, near Whittier, Calif., on April 18, 1952, at 3:34 a.m. PST. The crew of three and 26 passengers perished in the crash.

The trip was flown with a war-surplus Curtiss C-46F-1CU Commando, N8404C, msn 22466. The plane was delivered to the United States Army Air Force on July 23, 1945, as serial number 44-78643. Assigned to Manilla, Philippines, it did not depart the continental U.S. until April 19, 1946. The Commando was posted to Itazuke and Tachikawa, Japan, before its transfer to the Foreign Liquidation Commission on June 17, 1948. The Foreign Liquidation Commission sold the plane to Civil Air Transport, who registered it N8404C in December 1949. The plane was then sold to Air Charters Inc. On March 28, 1952, Robin Airlines leased the plane for a one-year period. Robin Airlines, a non-scheduled carrier, also was doing business under the name North Continent Airlines.

The airplane made one stop at Chicago, Ill., and then continued on to Kansas City, Mo., where the plane landed at 4:08 a.m. CST. A new crew boarded to take the flight west-

Top: *C-46F N8404C came to grief on the hills of La Puente, east of Los Angeles International Airport, the plane's final destination. The flight's crew of three and 26 passengers perished in the April 18, 1952 crash.*
(L. Griggers Collection via Macha)
Above: *War surplus Curtiss C-46s were plentiful after the war and formed the backbone of numerous non-scheduled carriers' fleets. The aircraft shown is the same type as N8404C.* (William T. Larkins)

ward to California. The crew consisted of Captain Lewis R. Powell, reserve Captain John D. Treher, co-pilot Charles K. Waldron, and hostess Dolores A. Bradford. Federal regulations required that a flight of the duration between Kansas City and Burbank required a third relief pilot — thus the inclusion of Captain Treher. He had dead-headed from his home in Chicago to meet the flight at Kansas City.

17

The morning after the crash: **above,** *the empennage and one of the R-2800-75 engines and its twisted Hamilton Standard propeller at the point of impact, 980 feet above sea level.* **Below:** *Rescue workers examine the wreckage. These are the largest pieces of fuselage that remained after the crash.* (Both L. Griggers Collection via Macha)

While preparing to depart Kansas City, the right oil cooler sprang a leak, requiring it to be replaced. This maintenance squawk delayed the flight more than 11 hours. The flight finally left Kansas City at 3:38 p.m. CST on a VFR flight plan with reserve Captain Treher flying left seat and Waldron sitting in the right seat. Captain Powell sat in the main cabin to relieve Treher on the next leg.

The flight landed at Wichita, Kansas, at 5:00 p.m. CST, to check the right engine's oil consumption. Thirty-four minutes later the flight departed for the West Coast. Treher was still at the controls when the flight arrived in the vicinity of Tucumcari, N.M., where they encountered severe thunderstorms. The plane turned east and flew to Amarillo, Texas, where it landed at 10:14 p.m. CST to wait out the weather. The plane's tanks were topped with 606 gallons of fuel and 32 quarts of oil.

Captain Treher left the flight and Captain Powell took command, departing at 11:02 p.m. CST headed for Phoenix, Arizona. The plane landed at Phoenix at 12:30 a.m. PST where three adults and one child deplaned. Flight 416W left Phoenix at 1:43 a.m. on a DVFR (Defense Visual Flight Rules) flight plan. It was instructed to cruise at 8,000 feet following Green 5 airway to Riverside, Calif., and then direct to the flight's terminus at Burbank.

Flight 416W requested the weather at both Los Angeles and Burbank at 3:13 a.m. PST. The flight was radioed the following information that was current as of 2:28 a.m.: "Burbank closed, visibility one-eighth mile, and Los Angeles measured 700 overcast, visibility two and one-quarter (miles) with haze and smoke, temperature 57 degrees, dew point 54, wind southeast (at) one (mph), altimeter 29.91." The pilots elected to file an IFR flight plan later in their journey. At 3:17 a.m., Flight 416W reported passing over the Riverside Range Station at 6,000 feet and requested an IFR approach to Los Angeles, giving an estimated time of 3:36 a.m. over Downey, California. Six minutes later, ARTC (Air Route Traffic Control)

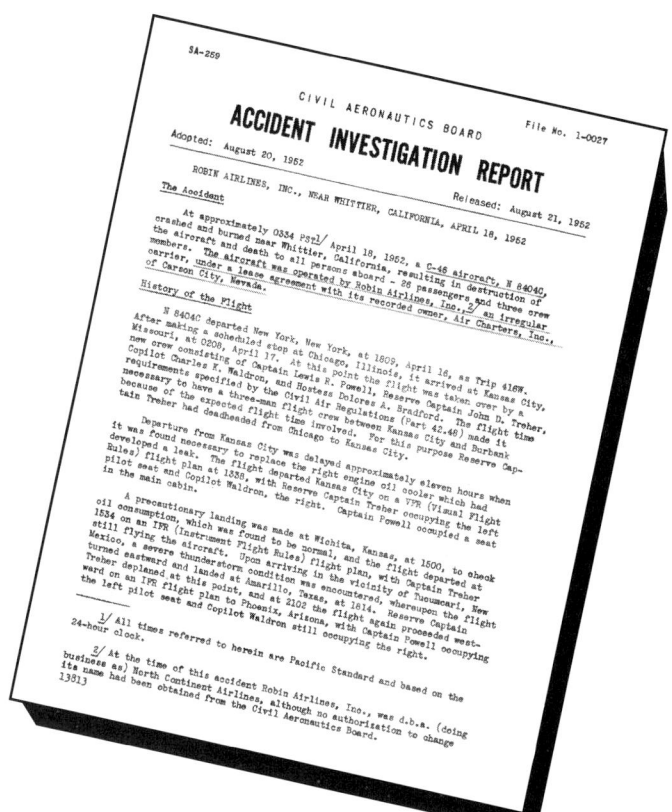

Civil Aeronautics Board Accident Investigation Report 1-0027, released Aug. 21, 1952, determined that the pilot descended too low to clear the terrain en route to Los Angeles International Airport. (Larkins Collection)

instructed the flight: "From present position to Downey radio beacon cruise at least 500 on top— Descend VFR and cross Downey and maintain 3,000—Contact Los Angeles Approach Control over La Habra—No delay expected." The pilots acknowledged the clearance and repeated the instructions.

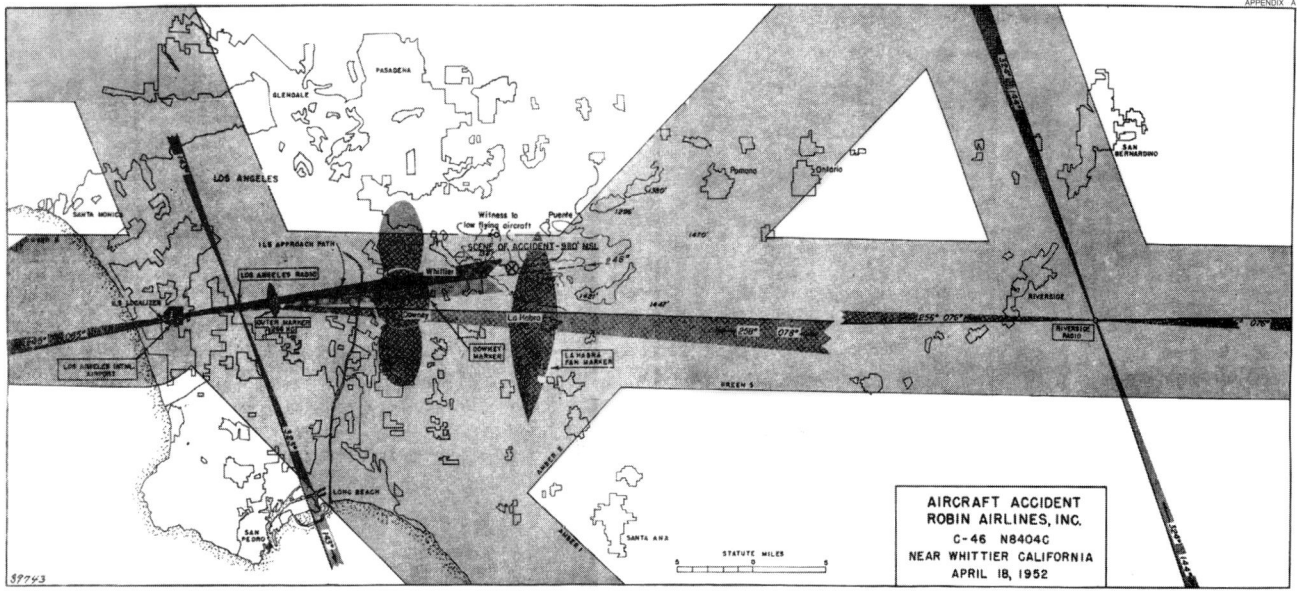

Airway map showing the flight path to Los Angeles International, the La Habra Fan Marker, location of the witnesses, and the crash site. Note that the crash site is just east of the published instrument landing approach path.
(Larkins Collection)

The massive tail of the C-46 remained intact while the fuselage was completely destroyed in the crash and subsequent fire. This wreckage was removed, but small pieces from the crash still are at the site. A metal detector is highly recommended.
(L. Griggers Collection via Macha)

The Final Four Minutes

The log of the Los Angeles Airport Chief Traffic Controller relates the last four minutes of Robin Airlines Flight 416W: "At 3:33 a.m. PST, N-04C called Los Angeles Approach Control on 119.9 mc (very loud and clear) advising he was in the vicinity of La Habra and was having difficulty with his ADF equipment because of static and that he would have to make some other type of approach. The approach controller asked N-04C if he had ILS equipment and whether he believed he could pick up the outer marker satisfactorily. The pilot answered in the affirmative and further stated that he could probably find the ILS Glide Path and proceed from there on in. N-04C was then cleared (3:34 a.m.) for a straight-in ILS approach from the Los Angeles Outer Marker and given the current Los Angeles weather and also advised that the Los Angeles L.F. Range was inoperative because it had been shut down for maintenance.

"The Los Angeles Approach Controller then immediately began watching for N-04C's appearance within the next few minutes on the Surveillance Radar Scope on the normal setting of 20-mile range as the flight continued westbound toward the Los Angeles Outer Marker. When no target appeared on the Surveillance Scope by 3:37 a.m. several radio calls were directed to N-04C by Los Angeles Approach Control. No response was received."

The Fate of Flight 416W

Hayden Jones, a Puente, Calif., rancher found the wreckage in the hills two miles east of Whittier and 22.5 miles east-northeast of LAX. A mass of twisted aluminum that once was Robin Airlines C-46 Flight 416W was scattered on the hillside at an altitude of 980 feet MSL and headed on a course of 260 degrees magnetic. The wreck site was 2,600 feet north of the ILS approach's center line, 7.5 miles east of the Downey fan marker.

The landing gear was down, and the flaps were fully retracted at the time of impact. The instrument panel clock was located. Its hour hand was loose on the spindal, the minute hand was stopped at 33 minutes after, and the second hand was frozen at 45 seconds. Examination of the area around the site revealed that the C-46 had struck the top of an adjacent hill, then continued to fly an additional 1,875 feet across a small canyon before impacting on the far side. The plane subsequently burned.

A local couple who lived 2.5 miles northwest of the crash site reported to officials that they had been disturbed from their sleep by a plane flying low overhead. Looking through a window, they saw the navigation lights of an airplane flying to the west. The plane flew out of sight, but not out of hearing range. The couple reported that the plane returned, now flying to the east, then out of hearing range. The Civil Aeronautics Board reported that Robin Airlines Flight 416W was the only airplane in the vicinity of the couple's home in the early morning hours of April 18.

Mitigating Factors

The pilot of Flight 416W, Captain Powell, was a very experienced aircraft commander. Powell had accumulated 7,913 hours flight time: 1,500 hours in the Douglas DC-3, 1,900 hours in the Douglas DC-4, and 600 hours in the Curtiss C-46. He held an Air Transport Pilot certificate that originally was issued in December 1945 and was reissued four months before the crash, on Dec. 23, 1951.

Powell had suffered a heart attack on March 30, 1951. He was sent to the hospital by ambulance where he remained until April 13, when he was sent home in an ambulance and confined to bed rest for the next six weeks. He gradually recovered. He remained under medical supervision until December 1951 and was last seen by his doctor on April 12, 1952.

During his recovery, Powell saw the physician who usually gave him his CAA medical evaluations, Dr. Francis C. Herzog. During his routine examination on Sept. 1, 1951, Dr. Herzog found no physical irregularities. When asked, Powell responded affirmatively that he had suffered a heart attack during the period since his last flight physical. According to federal regulations, Powell was asked to obtain and submit an electrocardiogram and a letter from his doctor for review by the CAA's regional medical officer.

The case was reviewed by the CAA Medical Standards Branch, Washington, D.C., who responded on Oct. 23, 1951: "...Mr. Powell, like others who have coronary artery disease, should not be certified for solo or pilot-in-command flying." Powell was issued a Class One medical certificate that was restricted to check pilot duties only.

Powell quickly took action in an attempt to convince the medical board that he should be allowed an unrestricted medical certificate. He was unsuccessful, and his restricted for-check-pilot-duties license stood. Powell was examined again by Dr. Herzog on March 14, 1952, to renew his restricted license. Dr. Herzog found no physical irregularities, but because of his heart history, reissued his restricted license.

A CAA aviation safety agent, G.O. Trapp, performed a ramp check on a Robin Airlines flight at Amarillo, Texas, on March 11, 1952. Powell was a crew member on this flight. Trapp stated in his report that Powell was not carrying the required airman identification card but, in a later statement, said that he did indeed examine Powell's medical certificate and noted no limitations. If the medical certificate was displayed in a wallet, the restrictions could have been folded under and not displayed.

Dr. Ellis stated during the post-crash investigation that the restrictions placed on Captain Powell were intended to restrict him to check-pilot duties only and that he was not to fly solo, or as any flight crew member—pilot-in-command or co-pilot.

The autopsy that was performed on Powell's body showed that his heart was severely damaged. Apparently it had recently suffered a hemorrhage but the post-mortem examination was unable to conclusively show whether the hemorrhage had occurred prior to the crash. The report stated that this hemorrhage may have "occurred within...probably a matter of hours, at the outside. It could have been immediately before, or it could have been a matter of several hours before (the crash)."

The Official Cause Of The Crash

After examining the wreckage and the aircraft's maintenance records, officials determined that there was no evidence of structural or mechanical failure occurring. The resulting fire showed there was fuel aboard. With this information, the CAB Accident Investigation Report dated Aug. 20,

Police, firemen, and officials from the medical examiner's office comb through the fuselage for clues to the accident's cause. Note the windscreen panels in the lower left of the photo. (L. Griggers Collection via Macha)

Above: *Photos from the day of the crash should help future Wreck Chasers locate the crash site. Note the fire truck on the road above and the wreck's proximity to the summit of the hill.* **Below**: *The gruesome task of removing the bodies goes on while investigators examine the wreckage. Note the main landing gear at the bottom of the ravine.* (L. Griggers Collection via Macha)

1952, stated "...it must be concluded that the cause of the accident was operational rather than mechanical. The board determines that the probable cause of this accident was the action of the pilot in voluntarily descending below the minimum altitude for which he was cleared and attempting an approach at an altitude too low to clear the terrain."

Powell's willful failure to comply with the restrictions placed on his medical certificate caused the deaths of himself, his two crew mates and 26 passengers.

The Crash Site

The plane went down in the La Puente Hills near Rose Hills Cemetery. The Los Angeles County Sherriff's Office lists the crash site as one-half mile south of the dead end of La Belle Street in the Whittier Hills district.

Although the major portions of the aircraft wreckage were removed shortly after the crash, it has been reported that small pieces of this aircraft can still be located. A metal detector is recommended.

WC

Fuel Transfer Starts In-Flight Fire
The Crash of United DC-6 NC37503

By Michael B. McComb

United DC-6 similar to the aircraft that crashed at Bryce Canyon. Fuel being transferred from one wing tank to another overflowed into the belly of the aircraft. When the cabin heater was switched on, an explosion and fire began. The pilots bravely attempted to reach the Bryce Canyon Airport, but time ran out for the crew of five and 46 passengers. (UPI/Bettmann)

Since the early days of powered flight, one of the greatest fears for any pilot has been an in-flight fire. Although less likely to happen in modern jet aircraft, the checklist procedure has remained relatively unchanged for more than 75 years. This serious and potentially devastating situation requires one of the quickest and most demanding responsive actions on the flight deck. Modern airline flight simulator sessions provide a number of in-flight fire emergencies such as engine fires, baggage compartment fires, and wheel well fires. Most aircraft fires today can be contained or extinguished by automatic fuel shutoff valves and halon-type fire bottles. These devices can be activated either thermally or manually by the flight crew. However, in the golden age of propliners, fire was fuel fed. Uncontained and uncontrolled, it could result in an extremely serious life or death situation. Such was the case Oct. 24, 1947, with United Airlines' Flight 608.

Shortly after 9 a.m., the crew of Flight 608 arrived via hotel courtesy car at the busy Los Angeles Airport terminal. After checking the en route and forecast weather, Captain E.L. McMillan was busy performing instrument and performance checks while First Officer G.G. Griesbach conducted the walk-around inspection. The aircraft, a Douglas DC-6, was one of United's newest to join the company's mainliner fleet. NC37510 had accumulated only 133 hours operation since its delivery date the previous March. The DC-6 was proudly named "Mainliner Seattle."

Its four Pratt & Whitney R-2800 engines ensured that it could climb higher and faster than its predecessor, the Douglas DC-4. The manufacturer had also installed the latest top-of-the-line Sperry autopilot system, and an advanced altitude rate controller for the cabin pressurization system. One major safety improvement was a fire detection and extinguishing system. Considered state-of-the-art for its time, the system could detect a fire in any of the four powerplants as well as both forward and aft cargo and hydraulic accessory compartments, and extinguish the fires with pressurized halon.

Cabin heating systems, on the other hand, remained mostly unchanged since the introduction of the DC-3 in the mid-1930s. The system operated very much like a modern jet engine. Aircraft fuel was fed into a burner can arrangement, while air was ducted via an air scoop on the belly of the aircraft. As the cabin temperature dropped, a thermostat closed a switch, igniting the fuel-air mixture in the burner can. The system worked well and is still being used on some light twin engine aircraft.

Given all the conveniences and technical innovations

Aerial views of the crash site. The above photo was taken Oct. 25, 1947, one day after the disaster. Highway 12 was then a dirt road. (UPI/Bettmann) *Today, lower, it has been paved and runs through the left third of the photo.* (Michael B. McComb)

The twisted tail section was the largest remaining piece of Mainliner Seattle. *Vertical fin appears to have severe fire damage.*
(UPI/Bettmann)

built into the DC-6, the engineers at Douglas had inadvertently developed and overlooked a sort of "Achilles Heel."

Mainliner Seattle To Chicago

Sunlight was beginning to break through the midmorning coastal layer as United began boarding Flight 608. Wearing their blue uniforms, stewardesses Shirley E. Brown, Helen F. Morrisey, and Sabina H. Joswich greeted passengers and directed them to their seats. Up front on the flight deck, McMillan was discussing a change of flight plan with Griesbach. They opted for a cruising altitude of 19,000 feet, which would give the aircraft a little less headwind and reduce the time drain along the usual seven-hour flight to Chicago.

Once all 46 passengers were aboard and their carry-ons placed in the open overhead compartments, Stewardess Brown closed the cabin door of the shiny silver and blue airliner for the last time. After receiving the "cabin secured" signal from the back, McMillan started each of the four gas-guzzling R-2800 engines. Family and friends at the gate watched as each engine roared to life, throwing clouds of oil smoke past the windows and obscuring the smiling faces of friends and loved ones.

Slowly the plane left the gate and taxied to Runway 25L. McMillan read the taxi checklist as Griesbach performed the functions — "hydraulic press, check…flaps, set…fuel…" Fuel was unbalanced with 2,400 pounds more in the left tank. However, a fuel imbalance was a common inconvenience in the airline industry, and the crew had no reason to suspect a safety concern. McMillan advised his copilot they would start a fuel transfer to the right tank upon reaching cruising altitude. Griesbach agreed, and they continued with the checklist.

Upon reaching the approach end of Runway 25L, and after performing engine run-up checks, Flight 608 was ready and cleared for takeoff. Griesbach eased the four power levers forward, and the DC-6 began its roll. McMillan called, "Four engines stable…manifold power set…60 knots…V1…V2…rotate…flight in less than 45s." Flight 608 lifted off, leaving behind a cloud of spray from the prior evening's light rain. Upon reaching a positive rate of climb, approximately 1,500 feet per minute, Griesbach called for "gear up" and McMillan complied.

McMillan had given Griesbach his leg to Chicago so Griesbach could add to his hours at the controls of the DC-6. Griesbach, 25, had been hired by United in 1945. He had accumulated more than 3,000 hours flying time, but only 66 hours in the DC-6.

McMillan, 42, had accumulate more than 15,000 hours flying time. He had joined United in 1936, and left during World War II to fly B-17s on bombing missions over Europe. When he returned to United in 1945, he flew DC-3s and DC-4s. Since being on line with the DC-6, he had accumulated 138 hours in the type.

Flight 608 had just passed Fontana, climbing through

United DC-6 NC37510 Mainliner Seattle came to rest on the far side of this barbed wire fence, just inside Bryce Canyon National Park. (Michael B. McComb)

10,000 feet, when they made a gradual left turn towards the Daggett navigation station. Below them were the San Bernardino mountains and to the north, the great expanse of California's Mojave Desert. As the flight progressed in a northeasterly direction, controllers with Los Angeles Center cleared Flight 608 to "climb and maintain 15,000 feet." McMillan acknowledged. The stewardesses had begun making their rounds, serving coffee and preparing lunch.

Passengers were settling down for the trip, some playing cards in the lounge while others, mostly business men, were catching up on work. The view outside the window was picture perfect. The scattered cumulus clouds seemed to glide by against the blue sky as the Douglas bore through the sky at more than four miles a minute.

Upon reaching their final cruising altitude of 19,000 feet, McMillan reached over and set cruise power. He began to crossfeed fuel from the left tank to the right tank, speeding up the process by placing the fuel transfer boost pumps on "HI." In this transfer configuration, the process would take less than 40 minutes.

At 11:40 a.m., Flight 608 reported its position over the gambling mecca of Las Vegas. In the distance on the right, passengers could see the still-forming Lake Mead and the western end of the Grand Canyon. Below lay the Charleston Range, where not more than five years before, actress Carole Lombard had lost her life in a plane crash.

The passengers settled in their seats for lunch and McMillan made an announcement over the P.A. system that, in a few minutes, they would be passing over the colorful butte and canyons of both Zion and Bryce national parks. For many on board, this was a highlight of the flight, and passengers began loading film into cameras for this and other sights along the way.

McMillan made his position report call with ARINC, "Salt Lake, United 608, Saint George, estimating Bryce Canyon 1222." It was shortly after this call that McMillan probably noticed that the right fuel tank had not only filled but overfilled. Overfilling was a common occurrence on Douglas aircraft because there were no automatic fuel crossfeed shutoff valves. Though it was something crews always tried to avoid, it would not have bothered either crew member. Normally, any extra fuel would just drain overboard.

But, unknown to McMillan and Griesbach, the highly volatile fuel was not quite leaving the plane in the 300 mph slipstream. Instead, it was flowing under the belly of the aircraft and directly into the cabin pressurization and heater air scoop at a rate of nearly five gallons per minute. Fuel had entered the compartment, soaking numerous articles of luggage and creating an extremely deadly situation.

Cabin Fire At 19,000 Feet

Meanwhile, oblivious to the situation below, passengers were busy taking pictures of the many brightly hued formations and the tall Ponderosa Pines of the Dixie National Forest. However, some passengers wanted to rest and complained to stewardesses that the cabin was cold. A stewardess told the flight crew, and McMillan made the adjustment on the thermostat. Just then a loud thud was heard and felt.

Many passengers in back thought the jolt was an air pocket. McMillan and Griesbach were puzzled and concerned but also thought it might have been turbulence. They were wrong. When the cabin heater came on, it caused a powerful fuel-air explosion, which completely blew the air scoop from the aircraft and ignited a raging fire underneath.

Soon McMillan and Griesbach were looking at an illuminated aft cargo compartment fire light and hearing the bell. Quickly, McMillan pulled the fire bottled handle, which silenced the bell but did little to put out the fire. At 1228, Griesbach put out a call to Salt Lake ARINC, "United 608, we have a fire in the aft cargo… unable… emergency landing Bryce… ." Smoke was beginning to fill the cabin as panic-stricken passengers began to scream. Some passengers in the back of the cabin may have noticed the floor becoming exceedingly warm, while others may have noticed the carpet beginning to smoke as the fire below began to burn through the floor.

McMillan, now at the helm, steeply descended toward the Bryce Canyon Airport, the controls shaking as the fire began to consume the elevators and rudder. As the plane flew over the town of Tropic, the main cabin door fell from the aircraft, taking with it a burning passenger. Other articles, such as luggage and portions of the aft galley also dropped from the aircraft as it made its fiery attempt to land.

When smoke began to fill the cockpit, McMillan and Griesbach were barely able to breath and donned oxygen masks as they strained to see the airport. The screams of panic previously heard in the cabin began to subside as each passenger became either consumed by fire or smoke.

Nearly six miles from landing, the aircraft was barely controllable, and Griesbach reported, "The tail is going out … we may get down and we may not."

At 12:26 p.m., the last radio transmission was received by Salt Lake controllers. "We may make it … approaching a strip." However, approximately one and one-half miles southwest of Bryce Canyon Airport, the Douglas' tail section gave way, shearing control cables. Flight 608 pitched into the plateau, sending burning wreckage and bodies flying for more than 1,500 feet.

Although the impact of the explosion could be heard for miles, no one saw the aircraft fall to earth. Charlie Francisco, now 78, was a boy living on a ranch near the town of Tropic, Utah, when he recalls seeing a large plane descending with flames and black smoke coming from it. He also recalls seeing burning fragments falling as it disappeared behind some hills. Then he heard an explosion. His account was like many others who witnessed the tragedy. When rescue workers reached the scene, there was little they could do. What was left of the aircraft was hauled off for scrap.

The Crash Site Today

It was the fall of 1992 when I arrived at Bryce Canyon and began my search for the crash site of Flight 608. Through previous research, I had a general idea of the location, but I did not have aerial photos, which made finding the site more difficult. According to the CAB report, the Douglas DC-6 crashed approximately one and one-half miles southeast of the Bryce Canyon Airport. I concentrated my efforts in this area.

I began a walking ground search. The first few hours were uneventful. However, I eventually crossed Highway 12 (the highway that runs through Tropic), and after climbing over a National Park boundary fence, I soon found myself standing amid the burned aluminum pieces of the once new propliner. The one thing that surprised me most about the site was the close proximity to a paved road. In fact, after further examination, I found that some pieces of wreckage had actually eroded from the hillside and were lying next to the highway. Unlike many of the other sites I have visited on which engines and large airframe structures remain, the Flight 608 site has none of these large pieces.

Most wreckage at the site measured approximately three to eight inches square, which is not unusual considering the proximity to a paved roadway. The variety of small pieces at the site includes everything from bits of window plexiglass to fragments of pale blue plastic UAL drinking cups. A closer examination may reveal pieces of jewelry, coins, flight instrument parts, and other small remnants from the plane.

Some of the old timers in the area told me how people used to look around the site and find wristwatches and rings. One story I heard occurred sometime between the late 1960s and early 1970s. A road crew was working in an area adjacent to the site across Highway 12 when one worker looked under a ledge and found a very weathered and torn Samsonite suitcase. Upon opening it, they found it contained the personal belongs of one of Flight 608's passengers. After hearing this story, I have no doubt that other items like this continue to lie amid the hills and cliffs of Bryce Canyon today.

However, since the site lies within the northern boundary of the Bryce Canyon National Park, all wreckage is considered federal property and may not be removed. To take anything is a felony. The Park Service does patrol the area, and fines can be very steep.

A visit to the Flight 608 site is a non-strenuous experience since there are no canyons to crawl down or mountains to climb. From Interstate 15 in Utah, take Highway 20 east to State Route 89, then turn right toward the town of Panguitch and continue south on 89. Make a left turn on Highway 12, pass the airport turnoff, and continue past the Highway 22 intersection and national park turnoff. In less than one-quarter mile, pull over at a large sign that reads, "Bryce Canyon National Park." You are there. (If you continue to drive and start to descend off the plateau, you have gone too far.)

Remember, the crash site is on the south side of the highway (the same side as the park sign). You must climb a four-foot barbed wire fence, but going over it is not illegal if you don't break it. The site lies a short walk further. Once again, don't remove anything you find.

WC

In his directions to the crash site, author Michael B. McComb states, "Make a left turn on Highway 12, pass the airport turnoff, and continue past the Highway 22 intersection and national park turnoff. In less than one-quarter mile, pull over at a large sign that reads, 'Bryce Canyon National Park.' You are there." (Michael B. McComb)

Collision Course: New York Mid-Air
The Tragic End Of Mainliner Will Rogers *And The* Star of Sicily

By Donald B. McComb Jr.

A stop sign on the corner of 7th Avenue and Sterling Place is dwarfed by the tail section of United Airlines' DC-8-11 N8013U, Mainliner Will Rogers. *The plane crashed into a residential section of Brooklyn following a mid-air collision with Trans World Airlines' Super Constellation* Star of Sicily. *(UPI/Bettmann)*

On June 30, 1956, a mid-air collision occurred over the Grand Canyon, in Arizona, involving a United Air Lines Douglas DC-7 and a Trans World Airlines Lockheed 1049 Constellation. Because of this accident, many improvements to the air traffic control system were implemented. While these changes were sweeping, they were not a panacea.

Nearly four and a half years later, a midair collision

The crash site in the intersection of 7th Avenue and Sterling Place bears no scars of the aviation tragedy that touched this peaceful corner of Brooklyn more than 30 years ago.
(Bill Hough)

occurred over New York City. As with the 1956 accident, the airlines involved were United and TWA, and in the case of TWA, the same aircraft type was involved. The combined death toll between the two aircraft came to 128 passengers and crew, which, believe it or not, was the same total for the 1956 accident. The New York accident also took the lives of six persons on the ground.

United Air Lines DC-8-11 N8013U, *Mainliner Will Rogers*, departed Los Angeles International Airport (LAX) on Dec. 16, 1960, at 12:20 a.m.—as Flight 856 en route to New York's Idlewild Airport (IDL). Flight 856, truly a "red-eye," thundered out over Santa Monica Bay into the darkness and turned east heading for Chicago O'Hare (ORD), its only stopover. N8013U, msn 45290/ln 22, had been delivered new to United on Dec. 22, 1959, and with just a few days short of its first year in service, had accumulated 2,429 hours of flight time.

The cockpit crew of Flight 856 was headed by Captain Robert Sawyer—who was assisted by First Officer Robert Fiebing and Second Officer Richard E. Pruitt. Captain Sawyer was 46 years old, and was employed by United Air Lines on Jan. 2, 1941. Type rated on the Boeing 247, DC-3, DC-4, DC-6, DC-7, and DC-8, Sawyer had logged a total of 19,100 hours. He passed his DC-8 flight check on June 6, 1960, and had accumulated 344 hours in the type as of Dec. 16, 1960.

First Officer Robert W. Fiebing was 40 years old and was employed by United Air Lines on May 1, 1951. He was rated in the DC-3, DC-4, DC-6, DC-7, and DC-8 aircraft and had accumulated a total of 8,400 hours, of which 416 were in DC-8 aircraft. Second Officer Richard E. Pruitt was 30 years old and joined United Air Lines on Sept. 15, 1955. He held both an Airline Transport Certificate and a Flight Engineer Certificate for the DC-6, DC-7, and DC-8. He had a total flight time of 8,500 hours, of which 379 hours were as flight engineer on DC-8 aircraft. If the crew's flight time in DC-8 aircraft seemed low, keep in mind that the DC-8 had only entered service in May 1959. Even the highest time DC-8 pilots in 1960 would have had no more then 1,000 hours by December 1960.

Flight 856 climbed rapidly to 33,000 feet (FL330) en route to Chicago. The plane arrived at 5:56 a.m. CST after an uneventful flight. After a two hour stop-over in Chicago the same flight crew that had brought the flight in from Los Angeles reboarded the aircraft. The flight attendants who had staffed the cabin from LAX departed the aircraft at ORD and were replaced by a fresh crew for the final leg of the trip into Idlewild as Flight 826. Fuel, food, and liquor were boarded and N8013U departed Chicago at 8:11 a.m. CST. Flight time to New York was estimated to be 1 hour and 29 minutes. The assigned cruising altitude of 27,000 feet (FL270) was reached at 9:36 a.m. On board the DC-8 making his first flight in a jet was Stephen Baltz age 11. Baltz had been seen off at Chicago by his father and was to be met at Idlewild by his mother and sister.

At 9:00 a.m. EST, eleven minutes before the United DC-8 left the ground at Chicago, a Trans World Airlines Lockheed 1049 Super Constellation N6907C *Star of Sicily* departed Port Columbus Airport, Columbus, Ohio, as Flight 266 for a 1 hour and 32 minute flight to New York's La Guardia Airport. N6907C, msn 4021, had been delivered new to Trans World Airlines on Aug. 11, 1952, and had been in service for eight years four months. N6907C had flown 21,555 hours. In its

TWA's Star of Sicily, *N6907C, a Lockheed 1049 Super Constellation, was delivered to the airline on Aug. 11, 1952. The Connie is shown here at an unidentified location midway through her career.* (Donald B. McComb Jr. Collection)

heyday N6907C had flown the world on TWA's extensive overseas routes. Subsequent to TWA placing 707s into service, the Connies were relegated to less prestigious routes such as Columbus to New York.

In command of Flight 266 was Captain David A. Wollam who was assisted by First Officer Dean T. Bowen, and Second Officer LeRoy L. Rosenthal. Captain Wollam was 39 years old. He joined Trans World Airlines on May 23, 1945, and was rated in the DC-3, Martin 202/404, and the Lockheed Constellation. Wollam had a total flight time of 14,583 hours, of which 267 were in the Constellation.

First Officer Dean T. Bowen was 32 years old and was employed by Trans World Airlines on July 13, 1953. He was rated in the 202/404 and the Constellation. Bowen had a total of 6,411 hours, of which 268 were in the Connie. Flight Engineer LeRoy L. Rosenthal was 30 years old and had been recently hired by TWA on Jan. 3, 1956. He held a current airframe and power plant certificate and a flight engineer certificate. Rosenthal qualified as a flight engineer with TWA on May 9, 1956, and had a total of 3,561 hours of which 204 were in Constellation aircraft.

After an uneventful flight, United 826 contacted New York Air Route Control Center at 10:12 a.m. The center answered "United 826, New York Center, roger, have your progress, radar service not available, descend to and maintain flight level 250 (25,000 feet)." Captain Sawyer reported leaving 27,000 feet at 10:14 a.m. One minute later, New York Center advised: "United 826 clearance limit is Preston intersection via Jet 60 Victor to Allentown, direct to Robbinsville, via Victor 123. Maintain flight level 250." United 826 acknowledged this message.

At 10:21 a.m. United 826 contacted ARINC (Aeronautical Radio, INCorporated), operator of United Air Lines' company communications system, and advised them that the "number two navigation receiver accessory unit is inoperative." This message was acknowledged by ARINC and reported to United Air Lines. As soon as Captain Sawyer had finished with this message, New York Center issued further clearance to descend to 13,000 feet. United 826 replied "we'd rather hold upstairs." At 10:22 a.m. New York Center called Flight 826, "United 826 New York Center, radar contact." United 826 replied, "Roger, we're cleared to 13,000 to maintain 25,000 until we had conversation with you. If we're going to have a delay we would rather hold upstairs than down. We're going to need three-quarters of a mile visibility, do you have the weather handy?" The center replied, "No, but I'll get it. There have been no delays until now."

Captain Sawyer reported over Allentown, Pennsylvania, at Flight Level 250 at 10:23 a.m. New York Center acknowledged, and at 10:24 a.m., New York Center advised that the Idlewild (now Kennedy International) weather was "...1,500 feet overcast, one-half mile, light rain, fog, altimeter 29.65."

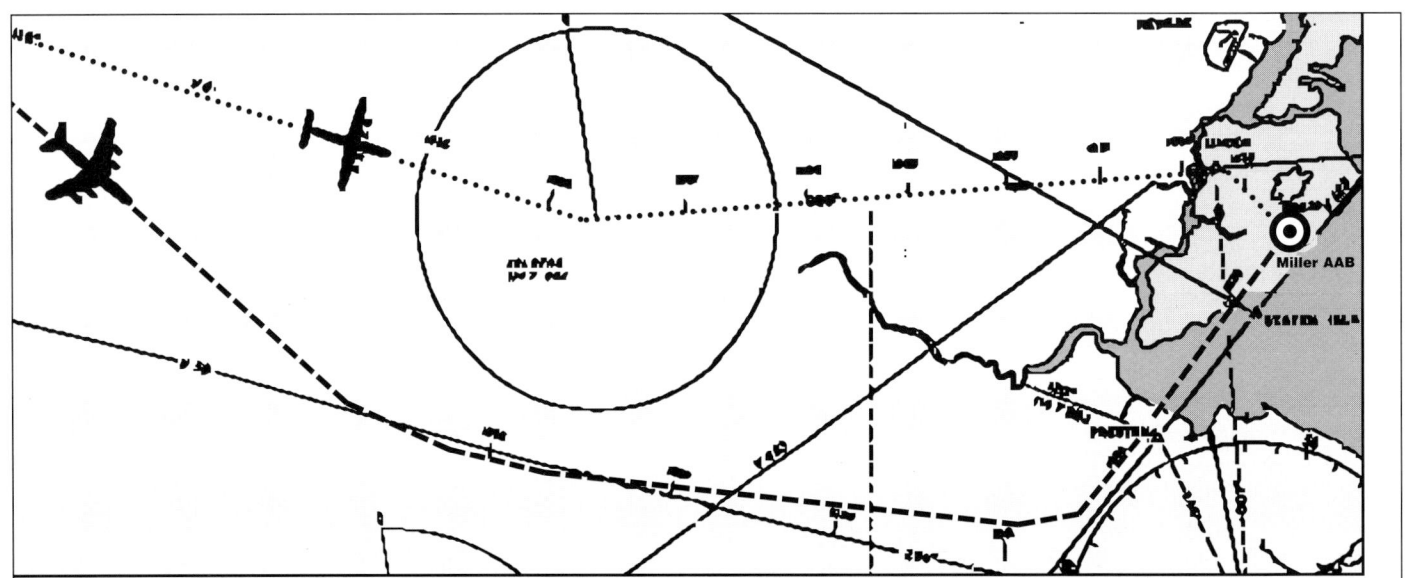

The Civil Aeronautics Board (CAB) issued this map showing the flight paths of the two airliners and the crash site on Staten Island. (CAB)

Accident investigators gather the wreckage of Star of Sicily *that crashed near Miller Field, Staten Island, New York. Horizontal stabilizer is on truck bed in foreground. Note rudder collapsed over on its side.* (UPI/Bettman)

LeRoy L. Rosenthal. Wollam was 39 years old, had joined Trans World Airlines on May 23, 1945, and was rated in the DC-3, Martin 202/404, and the Lockheed Constellation. He had a total flight time of 14,583 hours, of which 267 were in the Constellation.

First Officer Bowen, 32, was employed by Trans World Airlines on July 13, 1953. He was rated in the 202/404 and the Constellation, and had a total of 6,411 hours, of which 268 were in the Connie. Flight Engineer Rosenthal was 30 years old and had been recently hired by TWA on Jan. 3, 1956. He held a current airframe and power plant certificate and a flight engineer certificate. Rosenthal qualified as a flight engineer with TWA on May 9, 1956, and had a total of 3,561 hours, of which 204 were in Constellation aircraft.

After an uneventful flight, United 826 contacted New York Air Route Control Center at 10:12 a.m. The center answered, "United 826, New York Center, Roger. Have your progress, radar service not available. Descend to and maintain flight level 250 (25,000 feet)." Captain Sawyer reported leaving 27,000 feet at 10:14 a.m. One minute later, New York Center advised: "United 826 clearance limit is Preston intersection via Jet 60 Victor to Allentown, direct to Robbinsville, via Victor 123. Maintain flight level 250." United 826 acknowledged this message.

At 10:21 a.m. United 826 contacted ARINC (Aeronautical Radio INC.), operator of United Air Lines' company communications system, and advised them that the "number two navigation receiver accessory unit is inoperative." This message was acknowledged by ARINC and reported to United Air Lines. As soon as Captain Sawyer finished with this message, New York Center issued further clearance to descend to 13,000 feet. United 826 replied,

31

Firemen stand amid the wreckage of TWA Flight 266, Star of Sicily. *The Constellation broke into three pieces and came to rest in a Staten Island neighborhood.* (UPI/Bettmann)

"We'd rather hold upstairs." At 10:22 a.m. New York Center called Flight 826, "United 826 New York Center, radar contact." United 826 replied, "Roger, we're cleared to 13,000 to maintain 25,000 until we had conversation with you. If we're going to have a delay, we would rather hold upstairs than down. We're going to need three-quarters of a mile visibility; do you have the weather handy?" The center replied, "No, but I'll get it. There have been no delays until now."

At 10:23 a.m., Captain Sawyer reported over Allentown, Pennsylvania, at Flight Level 250. New York Center acknowledged, and at 10:24 a.m., advised that the Idlewild (now Kennedy International) weather was "...1,500 feet overcast, one-half mile, light rain, fog, altimeter 29.65." Shortly thereafter Captain Sawyer advised, "We're starting down."

In the rear of the DC-8's cabin, passengers noticed that the aircraft had started its descent and were busy turning in cups and other service items in preparation for landing. In the rear of the cabin Stephen Baltz watched as the cloud deck loomed closer and closer, finally engulfing the big jet in a sea of white.

Minutes earlier, at 10:19 a.m., the TWA Constellation reported to New York Center that it was "passing Allentown at 11,000 feet." Starting at 10:05 a.m., when the flight had been cruising at 19,000 feet, Captain Wollam had been descending in stages. New York Center advised that radar contact had been established, cleared the flight to Linden (New Jersey) Intersection, and requested the pilot to stand by for descent. At 10:21 a.m., New York Center further cleared TWA 266 to descend to and maintain 10,000 feet, then to 9,000 feet and to report leaving 10,000. TWA 266 acknowledged Center's call. At 10:27 a.m., TWA 266 advised that it was past the Solberg (New Jersey) VOR. The center acknowledged and advised that radar service was terminated, and to contact La Guardia Approach on 125.7. The crew acknowledged by repeating the frequency. At 10:29 a.m. Approach Control cleared the flight to descend to 6,000 feet. One minute later the controller advised TWA 266 to maintain present heading for a radar vector to the final approach course. At 10:30:49, Flight 266 was advised to reduce to approach speed. Less than two minutes later, the controller advised TWA 266 to turn right to a heading of 130 degrees. The crew acknowledged the transmission by repeating the heading. La Guardia Approach then cleared the flight to descend to 5,000 feet.

As the DC-8 flew through the heavy overcast, ARTCC (Air Route Traffic Control Center) amended the ATC clearance as follows, "826 cleared to proceed on to Victor 30 until intercepting Victor 123 and that way to Preston. It'll be a little bit quicker." This new routing shortened the distance to the Preston intersection by 11 miles. Because of this amended clearance, two things happened: Captain Sawyer had refused clearance to a lower altitude when it was first offered, so he now had to quickly lose a lot of attitude in a high speed descent. Furthermore, because one of the two VOR receivers on the DC-8 was inoperative, the crew had to use the one remaining to get a crossbearing with the VOR's at Colt's Neck, New Jersey, and Solberg, New Jersey, which would identify the Preston intersection. Because of the high

The DC-8's number four engine ingested a passenger from the Constellation during the collision. The engine broke away and fell into this playground of the General Berry houses in Dongan Hills, Staten Island. Patrolman William P. Pierce examines the twisted JT3C. (UPI/Bettmann)

speed descent and the limited time available, this would be very difficult to accomplish.

United 826 called Idlewild Approach Control at 10:33:28: "United 826 approaching Preston at 5,000 feet." This was the last communication from the flight. Flying at nearly 400 miles per hour, the DC-8 shot through the Preston intersection like a car running a red light. At 10:33:32, at an altitude of 5,200 feet over Staten Island, the DC-8's outer right wing and number four turbojet engine slammed into the cabin of the Constellation just aft of the right wing. Ripped from its mounts by the tremendous impact, the 6,000 pound Pratt and Whitney JT3C turbojet hurtled through the cabin of the Constellation, sucking a passenger into the turbine blades before flying out the other side of the fuselage. The DC-8 continued towards the triple tails of the Constellation, clipping off the right fin before the two aircraft separated. **WC**

The DC-3 That Crashed And Flew Again
The Crash Of American Airlines Douglas Sleeper Transport NC21752

Wearing her original name, Flagship Memphis, *American Airlines DST-217A NC21752, prepares to take on passengers at United Airport, Burbank, California. Note the interesting ground-handling equipment including a truck to air condition the cabin.*
(American Airlines via Davies)

By Ed Davies

Having identified an incident, the Wreck Chaser follows the parallel paths of researching the paper trail and pounding the real trail to the scene of the accident.

The first steps are determining the exact location of the crash site, obtaining right of entry, and scheduling the often grueling hike in. The reward is experiencing the unbelievable awe of seeing the field of wreckage.

The concurrent detective work needed to re-create the story of the crash may be prolonged, requiring patience and persistence. Using sources outlined in Volume I, *Wreck Chasing: A Guide To Finding Aircraft Crash Sites* and this edition, the Wreck Chaser can methodically piece together the paper trail and cement it by interviews with the people involved.

Scattered, charred fragments of twisted wreckage are the usual signature of an aircraft disaster. However, once in a while, a sturdy, very lucky aircraft survives complete destruction, is repaired and returns to the airways. The Wreck Chaser, in this case, may find the aircraft of his story still in operation at a major international airport or a quiet rural airfield.

Costly Boeing 747s have been classic survivors. A Japan Air Lines jumbo jet was blown off an icy taxi way at Anchorage in 1975. Eight months and 20,000 new parts later, a Boeing work crew handed it back to the airline. An American Airlines DC-10 suffered a structural failure that caused severe engine and hydraulic damage. The crew eventually got the crippled aircraft down at Detroit-Metropolitan Airport. There, an American Airlines maintenance crew from Tulsa rebuilt it in place and flew it out to serve with the airlines for many more years. A DC-8 landed in the water, short of San Francisco International Airport's Runway 28 Left. Hauled out of its watery wreck site, it was refurbished and returned to revenue service.

One of the most interesting examples of a survival event involved an American Airlines passenger DC-3. Crippled by a mid-air collision, it crash-landed in a snow-covered corn field. Leaving no trace for the Wreck Chaser, it was hauled from the wreck site, repaired, and in less than a month was back in scheduled service.

This particular story began when the author found a single-line entry in the airline's maintenance summary sheet for NC21752. It read: "12-24-44 Saline Mich. Collided with a C.A.A. plane while in flight." Few aircraft have survived a mid-air collision. Perhaps the story is worth telling of one that not only did, but is still flying half a century later

NC21752 was delivered new to American Airlines in August 1939 and was one of the last of 40 "Sleeper Transports" built by the Douglas Aircraft Company at Santa

Monica, California. The second row of tiny rectangular windows that served the upper berths distinguished this type from the regular 21-passenger DC-3. American Airlines used them on their southern transcontinental route, accommodating up to 14 overnight passengers.

The Davis-Douglas Company allocated the Manufacturer's Serial Number (msn) 100 to the Cloudster, the first aircraft they built. The next plane ordered was 101, and to this day Douglas still assigns unique consecutive identification numbers to its commercial aircraft. Though the registration may change many times during the life of the aircraft, the msn remains a constant invaluable identification item when tracing a plane's historic paper trail. DST-217A, NC21752's msn was 2165.

Pearl Harbor dramatically changed the U.S. airline industry, with many civilian planes being pressed into military service. The smaller passenger capacity DSTs were among the first to be released by the airlines. NC21752 had flown 7,956 hours, almost equivalent to non-stop flying day and night for a whole year, when it was purchased by the War Department in March 1942. Painted olive drab, with a "Star in Circle" insignia, it became a U.S. Army Air Corps type C-49E, serial number 42-43620.

Its wartime record, obtained from the U.S. Air Force Historical Research Center at Maxwell Air Force Base (See Volume I, *Wreck Chasing: A Guide To Finding Aircraft Crash Sites*), notes that this particular C-49E did not leave the continental United States. Based at Kansas City, Camp Douglas, Sedalia, Cheyenne, and Memphis, it was used for freight, passenger and medivac assignments.

Setting The Stage For A Mid-Air Collision

October 1944, American Airlines repurchased NC21752, converting it at their New York base to a standard 21-passenger DC-3. The distinguishing upper row of windows were plated over, and with 11,822 hours on the clock, it returned to civilian passenger service on Nov. 16, 1944. It was common practice at this time for individual aircraft to be named after cities on the airline's system. NC21752 became *Flagship San Antonio*.

Christmas Eve 1944, *Flagship San Antonio* was assigned to Flight 21, from New York to Chicago. The airliner departed Detroit City Airport, a scheduled intermediate stop, at 11:28 a.m. Eastern War Time. Piloting the flight was Captain Victor Evans, a nine-year veteran with American Airlines. The final leg of the flight was being flown by the First Officer J.R. Lyons. Stewardess Mary Brauer tended to the 15 passengers, many of them servicemen traveling from New York for Christmas furloughs. The youngest passenger was 2 year-old Candace Collet, traveling with her parents to their home in Chicago for the holiday. Other passengers included Elmer Johnson, electrician's mate of Portland, Oregon; Dr. E.G. Ahrens, Romeo, Michigan; Dr. R.S. Mitchell of Detroit; A.S. Tucker, seaman of Milwaukie, Oregon; Ensign R.D. Looney, Oklahoma City; Lt. R.I. Benson, Glenwood, Minnesota; apprentice seaman P.F. Walker, New Haven, Connecticut; Ensign R.F. Schemanski, Milwaukie, Oregon; Ensign W.W. Korn, Alameda California; Lt. A.O. D'Armour, Duluth, Minnesota; Ensign J.F. Garcia, Hayward, California; and Mr. G. Kasper, Detroit, Michigan.

The takeoff was routine, weather conditions good, and the aircraft proceeded on a westerly course at the 2,500-foot altitude assigned by the Detroit ATC. Stewardess Brauer served the box lunches to the relaxing passengers.

That same morning, at 11 a.m., a two-place Taylorcraft BL-65 fitted with skis and powered by a Lycoming 65 horsepower engine, took off from Ypsilanti Airport, Michigan. Registered NC24403 to Ypsilanti Air Service, it was based at the small airport 35 miles southwest of downtown Detroit. Aboard were Donald J. Gridley, flying instructor and part-time manager for Ypsilanti Air Service, and student pilot Miss

Flagship San Antonio on the ground at Saline, December 1944. Work had begun to remove the DST from the corn field for transport to Willow Run. One month later, the DST was back in revenue service.
(Davies Collection)

Eleanor Cramer. The training was associated with the Civil Air Patrol (C.A.P.). The notation on the American Airlines maintenance summary had wrongly described this as a C.A.A. flight. Cramer already had received five hours of flying instruction, and on this particular day was practicing left and right spins. Parachutes were mandatory for these maneuvers, and Cramer recalled that the chutes would make the small cockpit of the Taylorcraft even more restrictive. The flying lesson proceeded smoothly, and Cramer was flying the aircraft straight and level, about two miles south of Ypsilanti, positioning herself for a return landing.

Both aircraft were at the same altitude, each oblivious of the other's presence, as the faster DC-3 bore down on the small yellow trainer.

Researching The Crash

Perhaps this dramatic moment would be a good time to break, to review how the detailed information was gathered about an incident that took place 50 years ago.

Significant clues were the date and location of the incident. The airlines are quite naturally reticent about discussing accidents involving their equipment and thus are not a great source of information. However, knowing the date and location, I sent a letter to the National Transport Safety Board in Washington, D.C. It produced a copy of the "Accident Investigation Report." From it, I gleaned many further clues about the people, equipment and circumstances of the incident. Given the type, registration and msn, the F.A.A. in Oklahoma City, for a small fee, sent a copy of the microfiche containing the participating aircraft's registration and major maintenance history. Aircraft accidents are a newsworthy event, usually well documented by the local newspapers. Armed with the accident date, a call to the library in the city nearest the crash scene usually will generate an interested response, with perhaps photocopies from their newspaper microfilm files. In this case, I called the editor of the Saline *Reporter*. He in turn wrote a column on my request for information about the incident that generated several letters from citizens who recalled their participation and memories of that day.

Eventually, the meaningful sources of information began to peter out, and I hadn't found either pilot. It was time for the first barrel of the "Shotgun" research technique. I wrote to Arthur Pearcy, and he included the story as I then knew it, in the regular DC-3 column in the popular British aviation magazine *FlyPast*. Within a month, Pearcy sent me a copy of a letter he received from a *FlyPast* subscriber in the United States. It was the Taylorcraft pilot.

Gridley had retired and was living with his wife Lucy in Mackinaw City, Michigan. He still was in touch with his former student pilot, now Mrs. Eleanor Rose, who was living in San Diego, California. I corresponded with both, and they sent me photographs and photocopies of documents relating to the incident.

The Shotgun approach proved so successful, I decided to try the other barrel. Co-author Nicholas A. Veronico had just been appointed editor of the first regional edition of *Pacific Flyer*, a one-time experiment featuring Northern California and Bay Area aviation news. The story of the mid-air collision ran in the April 1988 issue under the title, "Indestructible DC-3." Bull's-eye again. A Saline, Michigan, reader sent a copy to George Hof, who at that time was vice president-Flight for American Airlines. He in turn forwarded it to retired American Airlines Captain J.R. Lyons, almost a neighbor of mine, living in nearby Hayward, California. I became good friends with Dick Lyons and his wife Gloria, and gathered first-hand his recollection of the mid-air incident. He was flying the DC-3 for the Detroit-to-Chicago leg. Pilot-in-command, Captain Vic Evans, had retired from the airlines many years earlier. He had passed away a year or so before I met Captain Lyons.

Diligent research of the paper work trail was now confirmed by the actual participants in the drama. They in turn were discovered by lucky hits from a Shotgun approach search for missing information — some of the many ways that the facts of yesterday's drama are slowly pieced together.

Impact! The DST And Taylorcraft Meet

To return to our particular story, two aircraft in a clear, empty sky were inexplicably drawing closer and closer together.

The first officer of the DC-3, leaning forward to tune in the auxiliary receiver for a range position check, saw a flash of yellow to the left and above. A split second later, he felt the impact, but could not recall whether he had intuitively taken any evasive action. The sound and jolt of the impact caused Captain Evans to take over the controls, though it took the combined efforts of both pilots to try to raise the left wing in an attempt to maintain level flight. The first officer used the radio to call the company to inform them of the situation and advise them that they probably would have to make an emergency landing about six minutes west of Romulus Army Air Base. Unable to maintain altitude, Captain Evans turned the aircraft a few degrees to the right for a wheels up, flaps down landing on a snow-covered corn field on the farm owned by Gilbert Dick. Stewardess Brauer cautioned all passengers to fasten their seat belts tightly and, immediately before the impact, the co-pilot cut all electrical power and pulled the mixture controls to the cut-off position. The crippled airliner plowed through three fences as it careened some 800 feet through the snow before coming to a silent halt.

Meanwhile, aboard the Taylorcraft, Gridley and his young student heard a shattering noise and lost all elevator and rudder control. One of the propellers of the overtaking DC-3 sheared off the right elevator and part of the vertical stabilizer. One of the skis from the Taylorcraft was embedded in the leading edge of the DC-3's left wing. For the first time they saw the airliner about a mile ahead and realized they had been involved in a dreaded mid-air collision. The small plane did not go out of control immediately, but Gridley, who realized that it was only a question of time, began helping Cramer unfasten her seat belt. He quickly coached her on the procedure, telling her to jump, count to 10, then pull the rip cord of her parachute. She remembers climbing out onto the strut, counting "One, seven, ten," pulling the rip cord and jumping. With the plane slowly turning north, Gridley jumped at about 1,600 feet. The doomed Taylorcraft plunged vertically to the ground. The tail was found alongside U.S. Highway 112; the remainder was strewn along a field south of the highway.

Tom and Blanche Bennet were startled by the roar of the wounded airliner as it passed low over their house on Willis Road. Donning their winter gear and high boots, Bennett and his father-in-law, Hartley Boutilier, struggled through the snow to where the DC-3 had come to rest. Despite the fact that airport neighborhood disaster drills hadn't been invented yet, the community public safety agencies sprang into action. A Mr. McBride at the Ford Bomber Plant called the Michigan State Police to inform them he had intercepted a radio distress transmission from the airliner. Patrol cars quickly honed in on the crash sites. Francis Lockwood, Saline's funeral director, whose hearse doubled as the local ambulance, raced to the scene.

Cramer floated earthward peacefully. She watched the DST crash land, her instructor's parachute open, and the Taylorcraft plunge into the fields below. Her landing was soft, and as she unbuckled her chute, a group of people came towards her. A young boy, who yelled with surprise, "It's a girl, a sissy," led the crowd. Cramer and Gridley ran towards each other, crying and hugging as they met. They made their way to the home of Edna Murray. From there, they were driven to the State Police Station, where they made their initial statements. Later Cramer went home to a delayed dinner and thankful prayers.

For a moment, silence reigned over the passenger plane as it lurched to a stop. The crew immediately ordered the passengers to evacuate the plane, fearing it might explode. Miraculously, none of the passengers or crew was hurt, nor had there been any panic during the frightening ordeal. Captain Evans led his passengers to the farmhouse, where they made their telephone calls and were supplied with food and coffee by Blanche Bennet and her neighbors. Co-pilot Lyons remained with the silent, wounded plane, guarding the mail and awaiting the authorities. U.S. Army doctors and ambulances arrived from Willow Run and the nearby Romulus Army Air Base, fortunately to find that their services were not required. The mail was off-loaded, and Bennet's nine-year old son Tom hauled it to the house on his sleigh. He carefully watched over it until the postal authorities came to pick it up. They put it on another flight from Willow Run. Tom was rewarded with five dollars for his diligence. A senior inspector from the CAA arrived from Dearborn, and Mr. Tanguay, air safety investigator from the CAB flew in, landing alongside the crippled airliner. They began taking statements for their report, which they hoped would identify the cause of the mid-air collision.

Close to 3 p.m., American Airlines' station manager from Detroit arrived to make arrangements for the stranded passengers. A Ford Motor Company bus took them first to Memorial Hospital in Ypsilanti for physical examinations. All were given a clean bill of health, but in the meantime weather intervened, canceling plans for a back-up plane to Chicago. Perhaps the passengers didn't want to fly again that day anyway. The evening ended with the airline hosting a Christmas Eve party for the crew and passengers at the Book-Cadillac Hotel.

State Police and the Civil Air Patrol maintained a guard over the downed airliner until Christmas Day, when engineers and mechanics from American Airlines arrived to start the recovery process. Engines and outer wing sections were removed. The DC-3 was jacked so the landing gear could be extended. During the evening of Dec. 26 and early morning of the next day, the fuselage was towed to the Willow Run Bomber Plant. Replacement engines and left wing were installed, enabling the aircraft to be ferried to the airline's maintenance base in New York. On Jan. 20, 1945, NC21752 was back in scheduled service with American Airlines.

With a mere 12,000 flying hours on the airframe, not a scratch on her refurbished fuselage, the rugged DC-3 could look forward to many more years of active flying. American eventually retired her in December 1948, placing the aircraft in storage at Tulsa, Oklahoma, until a buyer could be found. Purchased by Colonial Airlines in February 1949, the veteran

Engine cowls have been removed prior to separating the powerplant and accessory section from the firewall. During the collision, one of the Taylorcraft's skis stuck in the DST's port wing leading edge. **(Davies Collection)**

Douglas Sleeper Transport -217A, msn 2165, at Schellville/Sonoma Valley Airport, California, prior to the major restoration effort. The plane is currently owned by Dr. Lee Schaller, and it wears the paint scheme of Pacific National Airways. Schaller plans to restore this aircraft back to sleeper transport configuration. (Ed Davies)

DC-3 returned briefly to scheduled airline service before being sold again in August 1950.

Replaced in scheduled airline service by more modern equipment, many DC-3s were converted either for cargo operation or into luxurious corporate aircraft. N21752 was no exception, and in late 1950 it was transformed into corporate configuration by Remmert Werner at Lambert Field, St. Louis. A V.I.P. interior was fitted, and the original Wright Cyclone engines were replaced by the more available Pratt & Whitney Twin Wasp 14-cylinder radials. The new owner was Owens Illinois Glass Company of Toledo, Ohio.

During the early 1960s, the vintage DST was transformed yet again. Acquiring a high-density 24-seat interior, it became a "Flying Classroom" for Ohio State University at Columbus, Ohio. Re-registered N110SU, the plane was crewed by both faculty and staff pilots, with advanced students serving as copilots. Frequent uses included hauling university athletic teams (except the heavy football players) all over the Midwest, flying as an aerial platform for geography and geology studies, and for trips to New York with parties of theater students. Early in 1974, all three of the university's aging fleet of DC-3s were put up for sale.

Nevada Airlines of Las Vegas, operating scheduled passenger services over the Grand Canyon, took ownership in March 1975. Registration was changed once again, to N139D.

Pacific National Airways took possession during 1980, but soon afterwards the aircraft was withdrawn from service and stored in Las Vegas. Crocker National Bank offered N139D and her sistership N74KW for sale, and both were acquired by Dr. Lee Schaller, a San Francisco oral surgeon. Both were flown to Schellville/Sonoma Valley Airport in Northern California. N74KW was refurbished there, sold again, and ferried to new owners in Florida.

For the next 10 years, the historic DST, N139D, never flew. It occasionally was worked on, her colors fading in the hot California sun. It was considered as a possible candidate for flagship of the new C.R. Smith Museum built by American Airlines in Fort Worth, it was beaten out by a younger ex-American Airlines DC-3. Serious restoration began again in 1994. The work included a rebuilt interior and major work on the engines. As one of the oldest DC-3s in the country, one of the few originally built as a DST, and one of the very few to survive a mid-air collision, it soon may fly again.

History repeated itself on Nov. 27, 1951. An Eastern Air Lines DC-3, N25646, msn 2234, flying between Atlanta, Georgia, and St. Petersburg, Florida, was involved in a mid-air collision with an L-4J Piper Cub, again operated by the Civil Air Patrol. The larger aircraft made a normal landing at Taylor Field. The Piper Cub crashed into a heavily wooded area near Ocala, Florida, killing the pilot and only occupant.

WC

Final Destination: Grand Canyon MidAir
The Story of Flight 2 and Flight 718

By Donald B. McComb, Jr. with
Michael B. McComb

The Constellation involved in the Grand Canyon mid-air, N6902C Star of the Seine, *was similar to the 1049 shown at San Francisco above.* (William T. Larkins) Below: *United DC-7 N6324C* City of Vancouver *was similar to the aircraft shown above. At the time of the crash, N6324C had flown 5,115 hours since being delivered new to United.* (Michael B. McComb)

The grandfather of all midairs — the collision of Trans World Airlines Lockheed 1049 Super Constellation, N6902C, Star of the Seine, and United Airlines DC-7, N6324C, City of Vancouver, on June 30, 1956, shocked the world as the worst air disaster in the history of commercial aviation. The public outcry over this accident brought about a complete revamping of the air traffic control (ATC) system, creating new airways and providing radar separation nationwide.

Early on the morning of June 30, 1956, at Los Angeles International, the passengers preparing to board TWA's Flight 2 and United's Flight 718 were not thinking of radar separation or the ATC system. Undoubtedly, their thoughts concerned the start of vacations, a new job, or the countless other reasons people make cross-country flights. Those passengers that arrived at the airport early might have taken a moment to visit one of the airport concession stands for a pack of gum or a roll of Life Savers, never dreaming that a pack of candy would outlive them by 34 years.

Flight Into Eternity

Just after sunrise, TWA Flight 2, nonstop service for Kansas City, was being readied for departure. The aircraft was a Lockheed Constellation L-1049, N6902C. Delivered new to TWA on May 22, 1952, N6902C was the second A-model series aircraft delivered to TWA. The plane was named Star of the Seine for France's most famous river. After four years, one month, and one week of service, "02C" had flown 10,519 hours in TWA service. During the night, a line maintenance check had been performed, and the plane had been given a clean bill of health for the cross-country flight.

Several hundred feet away, United Airlines' Flight 718, nonstop service for Chicago, was in the final stages for departure. The aircraft for this flight was a Douglas DC-7,

39

Above: *This 1956 view of the Constellation's tail section shows the rugged, boulder strewn terrain of the crash site.* Below: *Additional view of the Constellation's tail section shows the rugged, boulder strewn terrain of the crash site.*
(Both Michael B. McComb Collection)

N6324C, which had been delivered new to United on Jan. 10, 1955, and carried the name City of Vancouver. In the one year, five months and 20 days with United, "24C" had flown 5,115 hours. The day before, N6324C had flown in from Idlewild as Flight 707, arriving at Los Angeles at 3:45 p.m. The DC-7 also had overnighted for a line maintenance check.

As the final boarding started, both aircraft were in the process of being fueled. The Connie was topped off with 3,300 gallons and the DC-7 with 3,850 gallons of high-octane gasoline. The weather en route was expected to pose no problem. The weather in Los Angeles that morning was typical for late June, coastal clouds and haze, that was expected to burn off by mid-morning.

The Star of the Seine 's doors were closed for the last time at 8:50 a.m. In command of Flight 2 was Captain Jack S. Gandy, Copilot James H. Ritner, and Flight Engineer Forrest D. Breyfogle. Captain Gandy had been with the company since 1939 and had a total of 14,922 hours, nearly half piloting Constellations. This trip would be his 178th between Los Angeles and Kansas City. One-by-one the big 2,700-horsepower Wright Cyclone R-3350 engines came to life, blowing clouds of oil smoke across the ramp.

While Flight 2 began its taxi out to Runway 25, United's Flight 718 also had fired up its engines and was making its way to Runway 25L. In command of the DC-7 was Captain Robert F. Shirley, Copilot Robert W. Harms, and Flight Engineer Gerard Fiore. Though Captain Shirley had fewer hours in DC-7s than Captain Gandy's time in Connies (1,238 vs. 7,208), he had accumulated 16,492 hours in his 19 years with United and had flown the Los Angeles/Chicago route on a regular basis since Oct. 1, 1955.

Captain Gandy held short at the end of Runway 25 as an American DC-7 touched down. As the American DC-7 cleared the runway, Captain Gandy received his takeoff clearance. He moved the Connie into position for takeoff. After a short pause, Gandy advanced the throttles, the four Wright R-3350's thundering roar drowning out all other sounds in the immediate area. At 9:01 a.m., the brakes were released, and the Star of the Seine began its final journey. N6902C lifted off easily, roaring over Dockweiller Beach and out over Santa Monica Bay.

The IFR flight plan filed by Captain Gandy called for a cruising altitude of 19,000 feet and a true airspeed of 270 knots. Flight 2 would fly to Kansas City via Green Airway 5, Amber Airway 2, crossing over Daggett direct Trinidad, direct Dodge City, and Victor Airway 10 into Kansas City.

While Flight 2 climbed out over Santa Monica Bay, Captain Shirley guided City of Vancouver into position for takeoff. Immediate clearance was given and the DC-7 rumbled off down the runway. At 9:04 a.m., three minutes after the TWA Connie broke ground, Captain Shirley eased back his control column, and the DC-7 sailed over the low hills that border the western boundary of Los Angeles International Airport.

Flight 718's flight plan would initially take the DC-7 on a more southerly route than that of Flight 2. Flight 718's was to cruise at 21,000 feet at a true airspeed of 288 knots via Green Airway 5, Palm Springs intersection, direct Painted Desert, direct Durango, direct Pueblo, direct St. Joseph, Victor Airway 116 Joliet, and Victor Airway 84 to Chicago-Midway Airport. However, once United 718 reached the Palm Springs intersection, its route would start to take a more northerly heading. The routes that TWA Flight 2 and United Flight 718 were flying would gradually come together over the Painted Desert. But with the United DC-7 flying 3,000 feet higher than the Connie, this should have posed no problem.

TWA Flight 2 had just reached its cruising altitude of 19,000 feet at 9:20 a.m. near Daggett, California. Captain Gandy, seeing thunderhead buildups on the far horizon called TWA Flight Operations and requested an altitude change on its IFR clearance from 19,000 feet to 21,000 feet and, if unable, 1,000 on top. The TWA radio operator who received the request called Los Angeles Air Route Traffic Control and at 9:21 a.m. advised, "TWA 2 is coming up on Daggett requesting 21,000 feet." The Los Angeles controller then contacted the Salt Lake ARTC controller and asked, "TWA 2 is requesting two one thousand, how does it look? I see he is Daggett direct Trinidad, I see you have United 718 crossing his altitude — in his way at two one thousand." Salt Lake replied, "Yes, their courses cross and they are right together."

The Los Angeles controller then called the TWA radio operator and said, "Advisory, TWA 2, unable to approve two one thousand." At this time, the radio operator interrupted and said, "Just a minute. I think he wants a thousand on top, yes a thousand on top until he can get it." After determining from the flight, through the TWA radio operator, that he was then 1,000 on top, the Los Angeles controller issued the following amended clearance: "ATC clears TWA 2, maintain at least 1,000 on top. Advise TWA 2 his traffic is United 718, direct Durango, estimating Needles at 9:57 a.m."

When TWA amended its flight plan from an assigned 19,000 feet to 1,000 feet on top, no information concerning this was given to United 718, nor was it required, even though the flights were in controlled airspace at the time. Neither aircraft was traffic for each other at the time TWA's amended clearance was granted, but in projecting their tracks eastward, both would cross Red Airway 15 with ill-defined horizontal separation.

Aboard both aircraft, people were settling down for the long flight ahead. The stewardesses on United 718 and the "hostesses" (as they were still called by TWA) began serving breakfast to their passengers. A few of the more observant passengers had already spotted the Colorado River up ahead. Beyond that lay Arizona and the sweep of the great plains.

TWA Flight 2 reported its position to company radio at Las Vegas at 9:59 a.m. Flight 2 had passed over Lake Mohave at 9:55 a.m., was 1,000 on top at 21,000 feet, and estimated reaching the Winslow omni station (Painted Desert) at 10:31 a.m., with Farmington the next reporting position.

At 9:58 a.m., United 718 made a position report to the CAA communication station at Needles, California Flight 718 reported that it was over Needles at 9:58 a.m. at 21,000 feet and estimated crossing the Painted Desert at 10:31 a.m., with Durango next. When the two aircraft crossed the Colorado River, they were about 40 air miles apart. With every mile, the distance between the Connie and DC-7 closed.

As they crossed over the Kaibab Plateau, the two aircraft were less than 10 miles apart. Some of the more keen eyed passengers on the DC-7 might have seen the white Connie,

Top: *Fuselage section of the TWA Constellation showing the red cheatline stripes.* Below: *Right forward fuselage section of the Constellation on Temple Butte.*

(Both Michael B. McComb Collection)

Above: *Main impact site of the* Star of the Seine. *Note the number of large pieces still at the crash site.* Below: *Memorial to those who perished in the June 30, 1956, Grand Canyon mid-air.* (Michael B. McComb Collection)

which would have appeared nearly motionless several miles away. One can also speculate that Flight 2 should have been visible to Captain Shirley from his left seat in the DC-7, but this was in the days before cockpit voice recorders. What was said on the flight decks of both aircraft before the accident will never be known.

The Final Seconds Of Flight 2 And Flight 718

Over the Grand Canyon that morning, towering cumulus buildups were forming — in some cases already climbing past 25,000 feet. As the two aircraft approached the eastern section of the canyon, they were probably maneuvering around clouds and, sometimes, punching through them. It is also possible they were banking to give their passengers a better view of the canyon.

What was happening in the last few seconds before impact will never be known. Were they in the clouds, in the clear, giving their passengers a better view of the canyon with a wing dipped, or steering around thunderheads?

What is known is: at 10:31 a.m. at a point just west of the confluence of the Colorado and Little Colorado rivers, the left aileron tip of the United DC-7 struck the leading edge of the center fin of the TWA Constellation. A split second later, the DC-7's lower surface of the left wing outboard of the Number One engine smashed into the upper aft fuselage of the Connie with a shattering impact. This started the breakup of the Connie's fuselage just forward of the pressure dome in the area of the rear galley.

The DC-7 continued to pass laterally over the top of the Connie, the left wing next making contact with the top of the left fin, shearing it off. Before the two aircraft separated — in less than one-half of a second, the propeller on the Number One engine of the DC-7 had inflicted a series of cuts in the aft baggage compartment of the Connie.

Minus its famous triple tail, the Connie pitched into a -8g dive. Passengers without their seatbelts fastened were hurled about the cabin to be pinned to the ceiling. Less than a minute later, the Star of the Seine smashed into the jagged rocks of a draw on the northeast slope of Temple Butte.

At impact, the Connie was in a slightly inverted dive going well over 400 mph. Totally fragmenting on ground contact, what little that was left of the aircraft was destroyed by the intense ground fire. Falling like a giant leaf, the massive triple tail/elevator assembly came to rest about 550 yards northeast of the main impact site. The ground impact knocked off both the left and right outboard fins from the tail assembly.

When the two aircraft separated after the collision, the United DC-7 was missing most of the left wing outboard of the Number One engine. Despite the fact that fuel was pouring from the ruptured wing tanks, there was no in-flight fire. The DC-7 began a turning descent to the left that was a little less rapid but no less fatal than TWA's.

It was at this time that a message was picked up by Aeronautical Radio (ARINC) in San Francisco and Salt Lake City. The message was not understood when it was received, but by playing it back was interpreted as: "Salt Lake, United 718 … ah … we're going in."

Could Captain Shirley have saved the DC-7? If he had cut power on the right wing Number Three and Four engines and increased power on Number Two, would he have been able to maintain some control? There are a couple of examples of airliners losing most of an outer wing (TWA 707 and PAA 707) and making safe landings. But both were jets and both incidents took place nearly 10 years later.

Some passengers on the left side of the DC-7, who just happened to be looking out the window, must certainly have had a "front row" seat to observe the collision as it occurred. Others seated on the left side, hearing the shouts and cries of fellow passengers were greeted with the sight of the twisted and stilled Number One propeller and fuel streaming from the torn wing. The rapidly increasing turning descent was an advance notice that the end was near.

As the DC-7 screamed downward, the aircraft shed bits and pieces of the elevators and rudder. As its speed approached Mach 1, the control surfaces encountered compressibility — where the airflow became greater than the control surfaces could act against. Captain Shirley had only one hope left. The faster the aircraft flew, the greater the lift generated, which could in theory pull up the nose of the aircraft into a climb that would slow the aircraft enough to break the sonic airflow grip on the control surfaces — provided there was enough airspace for this to occur.

Out the cockpit windows of the DC-7, Captain Shirley and First Officer Harms faced the imposing buttes and chasms of the Grand Canyon with plateaus nearly a mile above sea level. The collision occurring at 21,000 feet gave the DC-7 just over three miles of altitude to work with, and it looked like they might just make it. The rate of descent was slowing; the nose was starting to come up ever so slightly. For the crew of the DC-7, there was hope that they might just pull it off, just get the ship into a positive rate of climb to bleed off the excess airspeed. If they could get the plane under control, they could either head for an airport or make an emergency landing in the Painted Desert just beyond the Grand Canyon.

Another couple of thousand feet of altitude might have made the difference. But fate was not kind to United DC-7, N6324C, that day. Through the front cockpit windows, the crew watched with horror as the jagged sheer cliffs of a butte filled the windows. About a half second before impact, the DC-7 lost its left elevator due to the aerodynamic overload. With the nose down about 20 degrees and the left wing in a 10-degree bank, the DC-7 slammed into Chuar Butte just 10 feet below the top of the 4,050-foot ridge.

Most of the wreckage fell onto the ledges and into the 500-foot deep chimneys below the crash site. Some of the wreckage, such as the vertical fin, landing gear, and major parts of engines and propellers, came to rest on the south slope below the vertical butte face. Other major components, such as the horizontal stabilizers, elevator, and one engine, cleared the summit and came to rest on the northeast slope of the butte. Like the Connie, the ground fire was intense at the impact site, however, a great deal of the DC-7 wreckage fell away unburned.

Later that afternoon, one of the tour operators flying out of Grand Canyon Airport spotted smoke drifting up from the eastern part of the canyon. Having heard earlier in the day that two airliners were missing in the area, he decided to investigate. In a few minutes, he was flying over the two burned out crash sites. Confirmation came moments later when he spotted the Connie's tail section nearby.

The next day, rescue personnel from the Arizona

Top: *Aerial view of the United DC-7 crash site. Arrow indicates point of impact. This site was too remote to travel to in only four days.* Center: *Decaying TWA flight schedule recovered from the purse found at the Connie crash site.* Left: *R-3350 engine cylinder at the Constellation site.* (All Michael B. McComb)

43

National Guard made hazardous landings with helicopters at both crash sites. It was quickly determined that there was no hope for any of the passengers or crew from either aircraft. Rescue crews from the National Guard, U.S. Park Service, and a Swiss mountain climbing team were brought in to remove bodies and essential wreckage.

Based on the examination of the Connie's tail section, recovered sections of the Connie's fuselage and baggage door, and recovered sections from the DC-7's outer left wing (some of which was recovered at the TWA site), it did not take the investigators from the Civil Aeronautics Board long to find out that a collision had occurred. One of the end results of this investigation, in fact, led to the creation of the Federal Aviation Administration.

In the official report released by the CAB on April 17, 1957, the probable cause of the collision was as follows: "The board determines that the probable cause of the midair collision was that the pilots did not see each other in time to avoid the collision. It is not possible to determine why the pilots did not see each other, but the evidence suggests that it resulted from any one or a combination of the following factors: Intervening clouds reducing time for visual separation, visual limitations due to cockpit visibility, preoccupation with normal cockpit duties, preoccupation with matters unrelated to cockpit duties, such as attempting to provide the passengers with a more scenic view of the Grand Canyon area, physiological limits to human vision reducing the time opportunity to see and avoid the other aircraft, or insufficiency of en route air traffic advisory information due to inadequacy of facilities and lack of personnel in air traffic control."

The Crash Site Today

Over the years, the canyon has gradually reclaimed the crash sites. Rain and snow have washed and worn the fire blackened rocks to the point where you really have to know where to look in order to spot the sites. With the exception of most human remains and pieces of debris essential to the investigation, everything else was left where it had fallen. Since then, people on rafting trips down the Colorado have had the crash sites pointed out by their guides, and occasionally, sunlight reflects off pieces of the DC-7 high on the ledges of Chuar Butte.

In an effort to discourage souvenir hunters and to generally remove what some would call an eyesore, in 1976 and again in 1980, the Park Service contracted with the Arizona Air National Guard to remove most of the large items of wreckage. Large military helicopters were utilized in the removal process. Gone were the Connie's tail section and detached fins, landing gear, and large sections of the fuselage. The Connie's engines were blown up and most of the pieces removed. From the DC-7 site, only the large pieces that were projected down onto the lower ledges and slopes were removed in these operations. For the most part, more than 40 percent of the L-1049 remains on site, with more than 85 percent of the DC-7 still on Chuar Butte.

To a marine archaeologist, there are shipwrecks…and there are shipwrecks, like the Titanic or the Atocha. This accident is the aviation archaeologist's version of the Titanic. It is a very well-preserved site, and considering that aviation itself is less than a century old, it is a very old site. It has one other thing in common with the Titanic…it is very difficult to reach. Without a Jet Ranger III, reaching the sites meant hiking a roundtrip distance of just over 45 very hard miles, plus crossing rivers and running rapids.

Author Donald B. McComb was unable to hike to the crash site due to an injury suffered in a motorcycle accident. His involvement was in the planning, research, and documentation of all data and photographs obtained. Michael B. McComb was uniquely qualified to conduct this expedition. He holds an ATP certificate and, at the time, was a senior pilot for Air Grand Canyon. He flew daily over Temple and Chuar buttes. In addition, he is a trained archaeologist, having done work with the Bureau of Land Management on many important Indian archaeological sites. Donald B. McComb picks up the story:

While Mike and I had discussed going to the Connie and DC-7 crash sites for years, it was not until Mike started flying for Air Grand Canyon that planning for the trip started in earnest. As he put it, "Every day I flew what we called our Grand Tour. I could look down and see the reflections off the wreckage strewn across the slopes of Temple Butte and high on the cliffs of Chuar Butte. None of my passengers ever asked about the reflections on the rocks below. Of course, this is one bit of Grand Canyon history that I never mentioned unless asked."

Our initial plans called for a four-day trip, which included one day to reach the confluence of the Colorado and Little Colorado rivers. The second day would be spent crossing the river and setting up a camp near Chuar Butte. Since it was not expected to take very long to reach Chuar Butte, Mike planned to spend a good portion of the second afternoon examining the United site. The plan for day three was to get

Michael B. McComb near the Constellation crash site with his one-man raft used to travel the Colorado River. (Michael B. McComb)

up at daybreak and spend about six hours photographing and documenting finds. Camp would be broken around noon, and a short rafting trip down river would be made to a point near Temple Butte. Camp would be set up at once, and an examination would be made of the TWA site, photographing whatever wreckage was visible.

To enter the canyon in the first place, we had to obtain a backcountry permit, which works just like a flight plan and specifies a certain time period. In order to be out of the canyon by late afternoon on the fourth day, as required by the permit, Mike would have to make an early morning departure. This was vital, because permit holders who are not out of the canyon by the time specified are looked for and better have a good reason for being late.

This was a compressed schedule to be sure, but four days was all we had to work with at the time. Also, our main intention was to make this first trip an "inspection" tour to see if the sites were worthy of a longer expedition in the future. However, as we had discovered while investigating the Air West DC-9 crash site in the mountains above Los Angeles, trying to determine the difficulty in reaching a crash site and actually doing it are two different things. Our well organized four-day plan was doomed from the beginning.

Heading To The Crash Site

Before dawn on Oct. 26, 1990, Mike parked at Lippan Point near the head of the Tanner Trail, equipped with more than 60 pounds of gear in a backpack — cameras, food for a week, inflatable raft, emergency survival gear, including aerial flares, and just about everything you would rather have and not need, than need and not have. The first leg of the trek was a 12-mile hike to the end of Tanner Trail at the Colorado River.

The first few miles went fairly smoothly, then the trail just disappeared, becoming just a little less rocky than the rest of the terrain. The expected six-hour hike to the river took the whole day, and Camp One had to be set up next to the river at the end of Tanner Trail. The next morning, now several hours behind schedule, Mike set off down the Beamer Trail. He was climbing away from the river on a trail in no better condition that the Tanner Trail. About halfway along the Beamer Trail, Mike found the first evidence of the midair collision.

"The Beamer Trail portion of the trip, 12 miles in length, was much worse than I thought it would be. Even though I had topped off all my water bottles at the river, my water supply was going down much faster than expected. About halfway along the trail, I saw something reflecting sunlight some 50 feet off to the left. On closer inspection, I discovered what must have been part of the cabin interior trim from the fuselage of the Connie. This find consisted of a small piece of weathered plywood with an aluminum strip screwed along one edge. It measured about four feet long. This wreckage is consistent with some of the lighter debris that spilled out of the L-1049

Above: *Purse found at crash site held a TWA flight schedule, a roll of Life Savers, and coins, more than 30 years later.* Below: *Author Michael B. McComb holds a piece of skin at the Constellation crash site.* (Both Michael B. McComb)

after the collision. The sighting of this simple piece helped give me the added energy to continue on."

When Mike reached the end of the Beamer Trail, he had nearly exhausted his water supply and his daylight. He inflated his raft to cross the Little Colorado River, refilling the water bottles midstream, and setting up camp on the other side. At dawn the next morning, he crossed the Colorado River itself. The raft was beached for a few minutes while photos were shot of Chuar Butte. Because the windows were closing on the time he had available in the canyon, Mike decided to bypass the planned inspection of the United site. However, an examination through binoculars of the cliff face of Chuar Butte revealed several large pieces of DC-7 wreckage high on the ledges. On one of the pieces, the distinctive white and blue United paint scheme was still visible. Regretfully, the United site would have to wait.

Mike reboarded the raft for the short one-and-a-half mile trip down river, beaching on the shore near Temple Butte. This point was about 500 yards downhill from the Connie crash site. After the raft was unloaded, Mike nearly had a major disaster, courtesy of mother nature. "As I turned away from the now empty raft, a gust of wind came up. I turned back to the raft just in time to see it 20 feet in the air and nearly 15 feet out over the water. It had just touched down when I was halfway out to it. The water was freezing cold, but that raft was the only way out of the canyon. It was well anchored before I left it again."

Once camp was set up, Mike started for the impact site. There was no trail, only a very hard climb up a rugged slope. Just before reaching the main impact site itself, there is a narrow little canyon. It was here that first signs of the crash site were spotted. Pieces of what could have been parts of the wings or fuselage littered the ground. "I promptly dubbed the narrow canyon 'Connie Canyon.' When I reached the end of Connie Canyon, the main impact site spread out before me. Matching the terrain

45

to several photographs that were taken at the site in 1956, it was easy to see that the Connie had hit at the top of a small hill, spraying wreckage downhill towards the small canyon I had just come through."

For a couple of hundred feet in every direction, the ground was littered with pieces. Most were unidentifiable, but some were very easy to identify — crushed flight instruments, a massive cylinder from one of the R-3350 engines. Here and there were bent and twisted silverware with the TWA logo still visible; an exhaust stack from one of the engines. Coins littered the site, none with a date later than 1956. Human remains also were found and photographed. Sticking out of a pile of dirt and aircraft pieces, a purse was visible.

The purse was carefully extracted and opened. Inside were the remains of a badly decayed TWA flight schedule, a roll of Life Savers, a book of three-cent stamps, a pencil, lipstick, Max Factor makeup, $1.05 in change, and a Gulf Oil Company credit card. The name on the credit card matched a name on the passenger list.

Because this accident occurred in a National Park, nothing could be removed without a collection permit. However, our second canyon expedition, planned for late spring 1991, would be operated under the auspices of the U.S. Park Service and the Center for Aerospace Safety Education, Embry Riddle Aeronautical University in Prescott, Ariz. During that mission, selected items would be removed from the canyon for the U.S. Park Service collection of artifacts. But for this first trip, everything would remain at the site.

Because of the various delays in getting to the TWA site, Mike only had about three hours on site, not really enough time to do much. But it was enough time to decide that the

Articles found at the crash site include silverware, razor blades, seat belt buckles, and a travel clock.
(Harvey Gardner Collection)

TWA and the much harder to reach United sites were of prime interest for aviation archaeologists.

The next morning, after a rafting trip further down the river and through a couple of small rapids, that nearly swamped the raft, Mike crossed to the other side of the river to pick up the Beamer Trail for the long hike home. To sum up the whole experience, Mike said, "Throughout that long hike home, I made plans about what to do and what not to do on the next expedition into the canyon. But even more than that, my thoughts turned to the stark desolation and stillness of the crash site at Temple Butte. I think I now know how Ballard felt when he first looked down on the deck of the Titanic."

WC

Cross-Country To Disaster:
All On board United Flight 615 Perish

United DC-6B, identical to the plane that crashed in the hills behind Union City, California. (William T. Larkins)

By Nicholas A. Veronico

United Air Lines Flight 615, a Douglas DC-6B N37550, left Boston Massachusetts, at 5:32 p.m. EST on Aug. 23, 1951, en route to San Francisco. Three stops were made as the plane traveled westward; first at Hartford, Connecticut., next at Cleveland, Ohio, and finally at 9:59 p.m. CST, at Chicago, Illinois. During the layover at Chicago, Flight 615 changed crews.

When the flight departed Chicago at 10:59 p.m. CST, Captain Marion W. Hedden, 42, was at the controls. His First Officer was George A. Jewett, 35, assisted by Flight Engineer Marion A. Durante, 36, and Assistant Chief Flight Engineer Arthur W. Kessler, 43. Seeing to the passenger's needs were stewardesses Marilynn Murphy, 24, and La Verne Sholes, 22.

Captain Hedden began his career with United Air Lines on Nov. 1, 1939, and held an Airline Transport rating. He had accumulated 12,032 flight hours, including 417 in the DC-6 and 14 in the new DC-6B. He was granted a DC-6 rating on Jan. 15, 1951, and qualified on the DC-6B on April 26, 1951. The airplane Hedden was flying on the night of Aug. 23/24, was delivered new to United Air Lines on April 14, 1951. It had accumulated 361 hours total time without any engine changes or major mechanical problems.

Upon take-off from Chicago, Flight 615 carried 44 passengers, including two infants, and six crew members. The flight to California was filed as an instrument flight plan from Chicago to Oakland and then under visual flight rules across the bay to San Francisco. The flight plan's assigned altitude for the route was 18,000 feet, passing Denver, Colorado, and Milford, Utah, before heading direct to Oakland. The flight from Oakland to San Francisco was to be accomplished at an altitude ranging from 300 to 500 feet.

The cross-country flight was uneventful. All radio transmissions were routine and at 3:54 a.m. PST, Flight 615 was cleared to the Newark (California) fan marker with instructions to maintain 6,000 feet and to contact Oakland Approach Control over Altamont Pass, which separates the San Francisco Bay Area from California's fertile San Joaquin Valley. Although some 40 miles from the San Francisco Bay, the Altamont Pass area and its low-lying Livermore Valley are less than 15 miles from the Sacramento River Delta. During the summer months this area usually is blanketed by a marine layer with ground-hugging fog. When flying over the Altamont Pass, it would have been in clear air with a thick undercast.

47

Interesting aerial view of the crash site. Arrow shows path of airliner and ends pointing into Dry Creek Gulch. Vehicles are parked in center of saddle area. In the area near the large column of white smoke, center, a number of large pieces of wreckage were found 44 years after the crash.
(San Francisco Chronicle)

The flight reported over Stockton, California, at 4:11 a.m., at an altitude of 9,500 feet and descending. The flight was radioed the Oakland altimeter setting of 29.88 inches, which was acknowledged. Five minutes later, the flight reported passing the Altamont Intersection, and then contacted Oakland Approach Control for the first time. Approach Control cleared Flight 615 to the Oakland radio range station with instructions to remain no less than 500 feet above the cloud tops. United 615 then requested direct clearance to Newark with a straight-in range approach. The DC-6B reported that it was approaching the Hayward compass locator (between Altamont and Newark), and requested a straight-in Instrument Landing System (ILS) approach to Oakland. Flight 615 was instructed to stand-by for clearance until another aircraft in the area cleared. Captain Hedden then requested Oakland Approach Control to cancel his ILS approach request.

Bob Swanson found this engine mount on the far side of Dry Gulch, about 50 feet up from the bottom.
(Ian Abbott)

Oakland Approach Control's last instructions to the flight came at 4:25 a.m. The plane was cleared to fly from Newark on a straight-in approach on the southeast course of the Oakland radio range beacon. No further transmissions were received.

14.8 Miles Short Of Oakland

The *San Francisco Chronicle's* banner headline on the morning of Aug. 25, 1951, read, "Air Crash Kills 50: Big Liner Rams East Bay Hill." N37550 crashed 14.8 miles short of Oakland Airport in the hills behind Union City, California, at an altitude of 983 feet above sea level. The plane was on a course of 296 degrees, flying straight, and descending at the time of impact. A descent was confirmed by the fact that the hill it passed over prior to the crash was taller than the hill of impact. The DC-6B impacted, then cartwheeled over the peak and scattered itself across a small saddle and then into the canyon beyond. The wreckage field was 900 feet wide and 1,640 feet long. The fuel tanks exploded upon impact, causing a small grass fire.

The CAB report stated that "the main landing gear was extended at the time [of impact], and reasonable proof exists that the nose wheel was retracted, or nearly retracted. The main landing gear on this model extends before the nose gear and retracts after it. Wing flaps were between the fully retracted and 30 degrees extended position. All four engines were producing substantial power at the time of impact. Examination of propeller blade cuts in the earth and blade index settings showed that the blades were in the forward thrust range. Evidence indicated that the ground speed upon impact was between 225 and 240 miles per hour."

Weather conditions at the time of the crash were low broken stratus clouds with ceilings between 1,000 and 1,500 feet. Visibility below the clouds was better than six miles.

Crash investigators and firemen examine the remains of United's DC-6B on the saddle area behind Tolman Peak. The brush line in the lower right corner leads down into Dry Gulch Canyon. This area was pasture land and all debris on the grassy area have been removed. (San Francisco Chronicle)

Winds were reported below 10 knots and icing was not a factor as the freezing level was stated to be at 13,000 feet. The crew should have been able to see lights through the undercast from the towns of Niles, Centerville/Fremont, and Newark. The CAB went on to state that "the crash occurred during morning twilight and some light was also available from the moon. As Flight 615 broke out under the stratus at about 1,500 feet, downward visibility was possible, but ground objects and contours were probably difficult to recognize and identify. For this reason, it is believed that weather conditions were closer to Instrument Flight Rules (IFR) than VFR at the time of the accident.

"Three witnesses saw the flight in the vicinity of Newark. They stated that it was flying in and out of low clouds and noted nothing abnormal except that it was low. The impact site was shrouded in wisps of fog."

The Crash Site Today

Flight 615 crashed in the rising foothills on the eastern side of the San Francisco Bay near the borders of Hayward and Union City, California. The crash site is 14.8 miles from Oakland Airport on a true bearing of 123 degrees. The DC-6's final resting place is at the top of Dry Gulch Canyon on the eastern side of Tolman Peak.

In August 1951, the area was pasture land and the geographic features had not yet been named. The crash site is only 1.5 miles from Mission Boulevard, as the crow flies, at

Probable Flight Path map from Civil Aeronautics Board Accident Investigation Report SA-239, released March 12, 1952. Flight 615 approached from the east over the Altamont Pass en route to the Newark fan marker during the early morning hours of Aug. 24, 1951. After being cleared for an approach to Oakland, the plane struck coastal foothills east of Union City at 4:28 a.m. (William T. Larkins Collection)

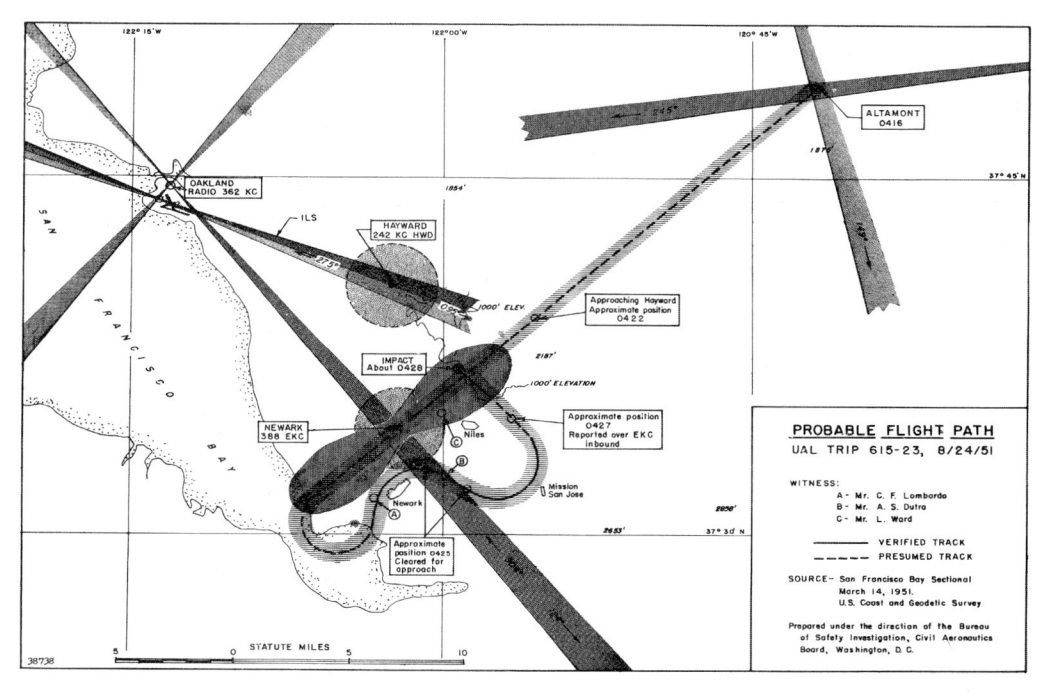

The crash site today: "A" shows the Tolman Peak Trail and the location of the Tolman Peak marker. "B" indicates the point where the emergency vehicles parked during the rescue and post-crash investigation. "C" is the area where the majority of the wreckage impacted after striking the hill from where this photo was taken. "D" is positioned just south of the Oakland Airport.
(Nicholas A. Veronico)

the southeastern end of Dry Creek Pioneer Regional Park. Dry Creek Pioneer and its adjoining Garin Regional Park are part of the extensive and well-managed East Bay Regional Park system. This system has preserved large amounts of open space and ridge tops in the East Bay hills.

Getting to this wreck, on a scale of one to 10 — 10 being hardest, is about a three. Depending on how adventurous you are once at the wreck site, going home can be a one (all downhill), or a seven if you decide to explore the wreckage that landed on the canyon sides of Dry Gulch.

Make sure to study the topo map for this crash site well in advance. Remember to carry the map. Two U.S. Geologic Survey 7.5 Minute Quadrangle maps will be required (see *Appendix IV, How To Obtain A Topo Map*). First obtain the Newark, California, quad for the walk into the wreck site. The crash site is located on the Niles, California, 7.5 Minute Quad in the upper left-hand corner at Tolman Peak.

Also highly recommended is a copy of the Garin & Dry Creek Pioneer Regional Parks guide map available from the East Bay Regional Park District. These maps can be

Seat pan found at eastern end of saddle. Numerous large pieces can be found after a dedicated search.
(Jim Rowan)

found in a stand at the park's entrance gate, but do not count on it. Obtain one prior to your trek by calling +1(510) 562-7275. This may cost about $1 plus postage.

Getting To The Site

Driving to the starting point of this Wreck Chase is relatively simple. From San Jose travel north on Highway 880, or from San Francisco/Oakland drive south on 880 to the Whipple Avenue offramp. Proceed east towards the hills until coming to Mission Boulevard. Turn left (north) and continue about one-half of a mile to Tamarack Drive. Turn right and follow the street to the end where you will come to the Dry Creek Pioneer Regional Park gate. Find a place to park on the street as there is no parking lot at this gate. This is an upscale neighborhood so your car should be safe.

Rattle Snakes, Poison Oak, And Scorpions: Getting To The Site And Back — Safely

Even though this crash site is "near civilization," proper precautions must be taken and common sense heeded. First, never hike alone. Wear long pants, a hat, and carry a long-sleeved shirt to avoid poison oak — lots of it! Carry enough water for the hike. Double what you consider to be an adequate amount of water. Bring a camera and, if you own or can rent one, a metal detector. Often just another heavy, useless piece of equipment lugged over Hell's half acre and never used, for this expedition it is highly recommended.

From the Tamarack Drive gate take the High Ridge Loop Trail for about half a mile. Then continue towards the right onto the Tolman Peak Trail. This will take you to Tolman Peak. It is a five-mile walk, relatively easy as hikes go. The trails are wide and well marked.

Once at the summit of Tolman Peak, continue to the saddle just behind it. The trail makes a 90-degree turn to the right at the saddle, leaving you facing a flat area with the brush filled canyon to the left and small peaks to the right.

The DC-6 impacted the peaks to the right and continued over to crash on the saddle in front of you. A number of large engine and firewall parts continued on into the canyon and still are there today.

Arriving at the saddle, my hiking companions and I agreed, "This is the place." Unable to locate any witnesses from the crash, we used aerial photos obtained from the *San Francisco Chronicle* to determine our position.

Splitting into three groups, Ed Davies gave the saddle a "once over" for any items visible through the tall grass. This would be a long shot as the property had been grazing land before becoming a park. The surface debris had been cleaned out years before, but it was worth a look. Ken Miller and Bob Swanson headed to the end of the saddle and proceeded to look down into Dry Gulch Canyon. Ian Abbott and I took the near and middle of the saddle and began to comb the brush on the rim of the canyon. Abbott was the first to locate debris, a small stringer with rivets on it, which confirmed the site.

Miller and Swanson came across a rattlesnake and an old encampment of some kind. The rattler was guarding a number of discarded cans, bottles, and ceramic dishes. While avoiding the snake, Miller noticed a shiny piece of metal on the far side of the canyon. Through binoculars the item looked round and spoked like a wheel. The group put off the decision to investigate the "wheel" until ready to depart the crash site. It would be a long slide down the near side

Veronico, left, and Ken Miller inspect a number of firewall forward parts and an engine magneto. These parts were located on the far side of Dry Gulch Canyon. (Ian Abbott)

and a real adventure to try and locate the wheel from the canyon bottom.

After searching the canyon side of the saddle, numerous small pieces of wreckage were found including an instrument case, stringers, and small sections of engine cylinders.

Once the decision to head back was made, our group headed, more accurately slid, down the hill into Dry Gulch. More parts were found including an oil tank and assorted rubber seals. At the canyon bottom, we split into two groups in an attempt to find the wheel. It was located without too much effort, but turned out to be a magneto. Other firewall forward parts including an engine mount and cowl flaps were found there.

Following the coutour of the canyon's bottom, the group was able to exit directly onto the Tolman Peak trail. This shaved about 30 minutes off the hike out.

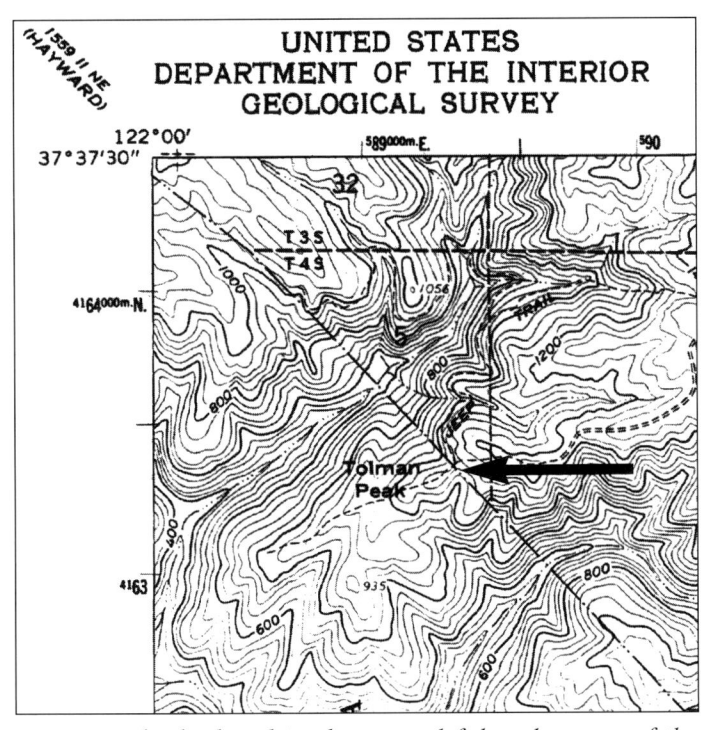

Postscript

The Aug. 28, 1951, issue of the *Oakland Tribune* reported that "...Fifty persons might be alive today if radar surveilance equipment had been functioning in the Oakland Municipal Airport control tower.." Installation of the equipment had been delayed by the Korean War as well as product setbacks. Radar would have instantly told controllers that the DC-6 was off course on its descent. This information could have been relayed to the crew, thus saving 50 lives.

WC

Prominently displayed in the upper left-hand corner of the Niles, California, 7.5 Minute Quad is Tolman Peak. United DC-6B N37550 came to rest east of the peak on a saddle (see arrow). This is a relatively easy site to reach on groomed East Bay Regional Park trails. (USGS)

Severe Turbulence Ends BOAC Flight 911 Over Mt. Fuji

By Donald B. McComb Jr

BOAC (British Overseas Airways Corporation) publicity shot of Boeing 707-436 G-APFE (msn 17706). It crashed March 5, 1966, on the slopes of Mt. Fuji. The crew of 11 plus 113 passengers perished in the disaster. (BOAC)

In the early afternoon of March 5, 1966, a BOAC (British Overseas Airways Corporation) Boeing 707-436 G-APFE sat patiently on the ramp at Tokyo International Airport awaiting service. On March 4, the 707 had flown from San Francisco to Honolulu and was scheduled to arrive in Tokyo late in the afternoon, where it would overnight. However, because of bad weather in Tokyo, the flight diverted to Itazuke, arriving at 6:00 p.m., local time. The flight, now at the midpoint of a round-the-world service that had originated in London, would continue on to Tokyo in the morning. Flight 911 was scheduled to arrive in Hong Kong that afternoon.

Seventy-five American tourists, who would board the flight in Tokyo, were on the trip of a lifetime. They were dealers and executives of Thermo King Corp. based in Minneapolis, Minnesota, and were finishing up a 14-day all-expenses paid trip that had been their reward for a sales campaign.

That night a Canadian Pacific DC-8-43 (CF-CPK) undershot the approach into Tokyo International Airport. CF-CPK's initial touchdown was 2,800 feet short of the runway threshold. It then hit a seawall, coming to rest near the end of the runway. The DC-8 exploded, killing 64 of the 72 on board.

Upon learning of the crash, there certainly was some apprehension among the passengers as they packed for the flight to Hong Kong, the last city on the tour.

Flight 911 took off from Itazuke at 11:25 a.m., local time, arriving at Tokyo's Haneda airport at 12:43 p.m. While the 707 was being serviced with fuel, food, and liquor, Captain Bernard Dobson and First Officer Malony received a weather briefing for the Tokyo-to-Hong Kong segment of the of the flight that was due to depart at 1:45 p.m. The weather briefing consisted, in part, of the following: "A depression intensified during the night and moved rapidly northeast across Japan. High pressure is developing over the Asian continent in conjunction with a low pressure area over the sea to the east of Japan. This is causing a steep pressure gradient from west to east over Japan at low levels." According to the observations taken at the Fuji-san weather station at the summit of Mt. Fuji, the wind was 60 to 70 knots out of the northwest and the temperature was 18 degrees Fahrenheit.

While the flight's crew of 11 and 113 passengers made their way out to the 707, someone paused for a moment to capture the scene with a camera. Taken from the observation deck, the photo shows G-APFE, its company colors of blue, white, and gold resplendent in the afternoon sunshine.

Mt. Fuji from the starboard side of a Boeing 707 similar to BOAC's F-APFE that crashed on March 5, 1966. Struck by a meteorological phenomenon known as a "mountain wave," this mass of air hit the 707 broadside with a force of 7.5gs.

(Author's Collection)

Surrounded by a sea of baggage carts, catering trucks and power carts, it looked well cared for. The scene must have inspired confidence in even the most nervous passenger. The 707 was one of the finest airliners around and BOAC was one of the safest airlines in the world.

Passengers and crew on board, G-APFE commenced communications with air traffic control at 1:42 p.m. First, permission was requested to start engines. Then the crew asked for clearance to amend the planned departure route asking instead for permission to initially climb in VMC (Visual Meteorological Conditions), undoubtedly to give the passengers a view of Mt. Fuji. This was approved and Flight 911 started its taxi to Runway 33L.

Flight 911 received takeoff clearance at 1:58 p.m. Captain Dobson advance the throttles and four Rolls Royce Conway 508s rapidly increased power output to 17,500 pounds of thrust each. G-APFE accelerated down the runway, lifting off and climbing out over Tokyo Bay. After making a wide right turn, Flight 911 flew south towards Yokohama and made a gradual turn southwest on a course that would take the 707 near Mt. Fuji. Approaching 17,000 feet, most of the passengers probably were starting to relax. Soon the drink cart would make its rounds through the cabin. One passenger seated on the right side of the aircraft over the wing took out his 8mm movie camera and started filming the breathtaking view outside. Investigators who later recovered this camera found footage of Tokyo airport, the Tanzawa Mountains, Lake Yamanaka, and after skipping two frames, something that looked like passenger seats and carpet.

When the 707 flew near the top of Mt. Fuji, it encountered a mountain wave condition. This phenomenon forms behind a mountain and flows at high speed down the lee side. Often these invisible waves are noticeable by the presence of a type of altocumulus cloud known as a lenticular cloud. Easy to spot, they are shaped like a giant lens with smooth edges. Today, however, no such clouds were present. This flow of air "broadsided" the 707 with a "wall of air" moving nearly 100 mph, from right to left across the path of the 707. Striking broadside, the aircraft was subjected to a 7.5g shock which snapped the vertical stabilizer off at its mounts — mounts that had been tested to take a 100,000 pound side load. The tail fin then fell, destroying the left horizontal stabilizer. The aircraft immediately pitched up and rapidly decelerated from 350 knots to less then 100 knots. It was at this point the the passenger filming lost control of the movie camera. Those standing or walking in the aisle were thrown into the cabin walls.

As G-APFE pitched up, all four engine pylons failed at the wing-to-pylon attach points. Simultaneously, the right wing failed outboard of the number-four engine and the rear fuselage failed at the aft main door. What was left of the 707 entered a flat spin falling nearly straight down, spilling fuel from fractured wing tanks. At around 6,000 feet, the fuselage forward of the wings, weakened by spin forces, broke away from the falling hulk. As this section failed, the center fuel tank fractured, flooding the forward baggage compartments and the electronics bays below the flight deck. This section went into a free fall and was destroyed and consumed by fire upon impact. Falling more slowly, the center section impacted in a forest at the base of Mt. Fuji. While this section did not burn, those alive after the in-flight breakup were killed by the ground impact.

Originally, investigators were puzzled by the fact that a modern jetliner had literally fallen apart on a clear day. Their attention quickly turned to the weather as a possible cause as mountain wave conditions were perfect that day. In addition, metallurgical tests showed fatigue cracks in the right attachment mount of the vertical stabilizer, though it could not be determined that these cracks had led to the structural failure. The best clue of all was the recovery of the 8mm

Blanket-covered bodies of crash victims lie on the ground at Gotemba, Japan, on March 5, 1966, after being recovered by police and troops from Japan's Self-Defense Forces. This was Japan's second major airline disaster in two days following the crash of a Canadian Pacific DC-8-43 CF-CPK. *(UPI/Bettmann)*

movie camera from the wreckage. Tests were performed on the camera to determine what amount of force would cause the camera to skip two frames and be pulled from the hands of the photographer. It was determined that a force of 7.5gs would cause the camera to skip two frames.

From an investigator's standpoint, there was not much to work with. The aircraft was not equipped with a cockpit voice recorder and the flight data recorder, primitive by today's standards but state-of-the-art in 1966, was mounted in the aircraft's nose below the cockpit floor. It was completely destroyed by the post-crash fire that also consumed the aircraft's instrumentation. In addition, a U.S. Navy Skyhawk which overflew the site of the accident shortly after it occurred, encountered turbulence so severe that the aircraft was grounded for inspection after landing. The official report as issued by the Japanese government listed the possible cause of the accident as follows: "The probable cause of the accident is that the aircraft suddenly encountered abnormally severe turbulence over Gotemba City, which imposed a gust load considerably in excess of the design limit."

In terms of Wreck Chasing, getting to this accident site should be simple. Do your homework thoroughly. Then hop a flight to Tokyo, rent a car and drive to Gotemba City. There, in the shadow of Mt. Fuji, ask the locals about the air disaster that happened in 1966. This part might take some time, language problems and all, but once you get proper directions you should be able to pinpoint the site. Is there anything left on the site? Most certainly you should be able to find something, even if it's only a small piece of metal with a couple of rivets. Experience shows that rarely is an accident site totally cleaned up. The exception to this is a crash that occurs in a city or on an airport. Recently, I received a phone call from a reader who informed me that while stationed in Japan back in 1969-'70, as a crewman on helicopters, he had overflown the crash site several times. Cruising over the eastern slopes of Mt. Fuji, he spotted large pieces of the 707. Its blue-and-white paint was still visible.

Is it still there? Who knows. All I can think is that if there was no motivation to remove the wreckage between 1966 and 1970, that it is still possible for it to be there. Good luck!, and please let me know what you find.

WC

Lightning Strike Ignites Explosion — Downing Pan Am Flight 214

By Donald B. McComb Jr.

Boeing 707-121 N709PA (msn 17588) was lost Dec. 8, 1963, over Elkton, Maryland, when a lightning strike ignited fuel fumes in the left wing. (Author's Collection)

Sunday morning, Dec. 8, 1963, dawned no different then any other at Philadelphia International Airport. Among the many airliners being made ready for flight was Pan American Boeing 707-121 N709PA (msn 17588). It would operate Flight 213, service to San Juan, Puerto Rico, with an en route stop at Baltimore. After spending a couple of hours on the ground at San Juan, the 707 would return as Flight 214, stopping once again in Baltimore before arriving back at Philadelphia at 8:30 p.m. local time. Truly an historic aircraft, N709PA was the third-production 707, and on Aug. 15, 1958, became the first jetliner delivered to a U.S. air carrier. After five years and four months in service, N709PA had flown 14,602 hours as of Dec. 8.

Only one incident had marred an otherwise trouble-free flying career. During a training flight on Feb. 25, 1959, while flying near its minimum controlable air speed at 8,000 feet over France, N709PA was allowed to stall. During the recovery, one of the engines and most of the pylon separated from the aircraft landing in a farmer's field. The plane continued on to Heathrow, where it was inspected and repaired.

Flight 213 departed Philadelphia prior to 10 a.m. local time. In command was Captain George F. Knuth, 45 years old, a veteran Pan American pilot with 17,049 hours, 2,890 of which were in the 707. In the right seat was First Officer John R. Dale, employed since 1955, who had 13,963 hours, of which 2,681 were in type. Second Officer Paul L. Orringer was the navigator, but also was rated as a pilot on the 707. He had logged 10,008 hours, of which 2,808 were in 707s. John R. Kantlehner served as flight engineer. He had flown 6,066 hours, of which 76 were in the 707. The flight to San Juan was without incident and arrived shortly after 2 p.m. local time. While on the ground, the captain was briefed on the weather along his intended route back to Philadelphia. The crew was provided with a flight folder containing the required weather documents. While the crew was being briefed, N709PA was being prepared for the return flight. Flight 213 landed with 25,500 pounds of Jet-A fuel remaining aboard and took on an additional 52,500 pounds of Jet-B fuel. The difference between the two fuels is that Jet-A is standard aviation kerosene, not much different then what is used in kerosene heaters, and Jet-B is a wide-cut gasoline-type fuel with a very low flash point and a maximum vapor pressure of 3 pounds per square inch. When the fueling was completed, the wing tanks of the 707 contained between 64 and 69 percent Jet-B fuel, while the center tank, which on the 707 holds 7,306 gallons of fuel, was 100 percent Jet-B.

At 4:10 p.m. local time, the 707 departed San Juan as Flight 214. The flight north was uneventful, the 707 skirting the eastern coast of the United States with the late-afternoon sunshine turning to twilight. At 7:35 p.m., *Clipper Tradewind* touched down at Baltimore's Friendship International Airport. Seventy-one passengers deplaned and none were boarded for the short flight to Philadelphia. At 8:25 p.m., N709PA thundered off into the night for the short 20-minute flight. As the aircraft climbed out, Baltimore Departure Control provided radar vectors to airway Victor 44 where control of the flight was passed to New Castle Approach Control. Communication was established and the 707 was cleared to descend to 5,000 feet, then to the New Castle VOR. Captain Knuth reported over the VOR at 8:42 p.m., and was transferred to Philadelphia Approach Control that advised, "Philadelphia weather now 700 scattered, measured 800 broken, 1,000 overcast, six miles visibility with rain showers. Altimeter 29.45. The surface wind is 280 degrees at 20 knots, with gusts to 30 knots. I've got five aircraft holding until the extreme winds have passed...do you wish to be cleared for an approach, or would you like to hold until the squall line...passes Philadelphia?"

The crew advised approach control that they would hold. Flight 214 was instructed to hold west of the New Castle VOR on the 270 degree radial and was given an expected clearance time of 9:10 p.m. The crew requested and received permission to use two-minute legs in the holding pattern. Flight 214 advised that that they were ready to start the approach at 8:50 p.m. Approach control instructed them to continue holding and they would be cleared as soon as possible. Flight 214 acknowledged with, "Roger, no hurry, just wanted to let you know that...we'll accept a clearance."

For the next eight minutes, Flight 214 flew through the rainy night in a holding pattern near the small community of Elkton, Maryland. Also in the holding pattern, but 1,000 feet higher, was National Airlines Flight 16, operated by N875C, a Douglas DC-8-51. At 8:58 p.m., the following transmission was heard on Philadelphia Approach Control frequency. "MAYDAY, MAYDAY, MAYDAY, Clipper 214 out of control...here we go." Overhead the first officer of National 16 watched in horror as the Boeing 707 spun out of control. "Clipper 214 is going down in flames," he radioed tersely. One thousand feet below, a lightning bolt had struck the left wing of the 707, resulting in an explosion that caused the wing to fail between the Number One and Two engines. Flight 214 began spinning earthward trailing a streamer of fire. At nearly 500 mph, it fell into a cornfield near Red Hill, two and one-half miles northeast of Elkton. The main wreckage missed a farmhouse by 100 feet. While the majority of the wreckage was located at the main impact point, parts were strewn as far away as 19,600 feet.

At first, crash investigators were hard pressed to believe that lightning indeed had caused the demise of Flight 214. Only once before in the history of commercial aviation was it proven that lightning had caused an accident. This was the loss of a TWA 1649 Constellation near Milan, Italy, in 1959.

After an extensive investigation it was determined that two factors were instrumental in the cause of this accident. Fuel type and lack of static wicks. It was determined that the outer left wing tank was empty. The lack of static wicks allowed the static charge on the aircraft to build up as the aircraft flew in and out of the clouds. Recovered wreckage showed a lightning strike on the left wing tip and on the tip of the high-frequency antenna probe located at the top of the 707's vertical stabilizer. Whether it was the lightning strike itself that caused the explosion, or the static build-up, will never be known for sure. However, gasoline vapors in the left surge tank did ignite, causing a greater explosion in the

The cockpit crew of National Airlines Flight 16, a Douglas DC-8-51 N875C (msn 45635), witnessed the death plunge of Pan Am Flight 214 on the night of Dec. 8, 1963. (ATP/Airliners America)

left outer wing tank. At the time of the accident, N709PA was one of ninety-seven 707's flying that had not had the static wicks installed. The end result was that the use of gasoline type turbine fuels became almost obsolete in commercial aviation, and static wicks became standard equipment on every airliner in service.

Notes for Wreck Chasers: In December 1990, I attempted to locate this accident site. This was a spur of the moment trip. At the time I was visiting a friend in Glen Burnie, Maryland, and was armed only with photocopies of the accident. My goal was to take a look at the town itself and ask around to see if someone could point me in the right direction. Elkton is located about an hour north of Baltimore near the Delaware border. Once in Elkton, I stopped at a small store. The clerk remembered the accident and gave general directions that took me three miles east of town. (Note, if you cross into Delaware you've gone too far!) I located a cornfield next to a farmhouse that matched the general description of the crash site. This location also matched the store clerk's description. Unfortunately, it was getting dark and there was no time to pursue things further. Satisfied that I had at least found the general area of the accident, if not the actual site, I prepared to return to Baltimore. As a general observation, everything in Elkton looks like it has not changed in 40 years. In all likelihood, the impact site still is a cornfield so there is some potential for wreck chasing. Since this area is predominently farm land, written permission to enter fields should be obtained and carried when searching for this site.

One last note of amusement: On the way out of town I stopped at a payphone. Using the photocopies, I checked the local listings for the name of a resident whose house was almost struck by a large piece of wreckage. Finding the name, I dialed the number, introduced myself, and explained that I was searching for the point of impact. I asked if this person would be open to a short interview. After hearing me out, the gentleman said, "I'm not interested!" and hung up. Potential Wreck Chasers should keep this in mind if you decide to use this approach to do your research.

wc

"I Just Happened To Look Up... I Saw It Instantly..."

Air West DC-9 and Marine Corps F-4B Phantom Collide Near L.A.

By Nicholas A. Veronico and Donald B. McComb Jr.

Air West DC-9-31 N9345 (msn 47441) taxies out for take off at Los Angeles International Airport (LAX) during its brief career. The plane was delivered new to Air West on June 4, 1969, and collided with a Marine Corps F-4B Phantom II on June 6, 1971. Forty-four passengers and five crew on board the DC-9 as well as the pilot of the F-4 perished in the accident. Only the Phantom's rear seat Radar Intercept Officer survived. (ATP/Airliners America)

Among the many airliners preparing to depart at Los Angeles International Airport (LAX) in the late afternoon of June 6, 1971, was Air West Flight 706. It was Air West's daily service from Los Angeles to Seattle with stops at Salt Lake City, Utah; Boise and Lewiston, Idaho; and Pasco and Yakima, Washington.

Equipment for Flight 706 was N9345, a DC-9-31 (msn 47441). Delivered new to Air West on June 4, 1969, the 503rd DC-9 off the line at Long Beach, it had flown 5,542 hours with Air West.

In command was Captain Theodore Nicolay and First Officer Price Bruner. Between them, the Seattle-based flight crew had nearly 30,000 hours of flying experience. At 6:02 p.m. local time, N9345 departed LAX with 44 passengers and a crew of five. Taking off to the west into the setting sun, the DC-9 headed out over Santa Monica Bay for approximately five miles before making a right turn to heading 040 degrees. This course would take the DC-9 over Santa Monica, Beverly Hills, and just to the north of Dodger Stadium, before heading out over the rugged San Gabriel Mountains.

While the DC-9 was climbing out from LAX, a Marine Corps McDonnell Douglas F-4B Phantom II (Bureau Number 151458) was passing over Palmdale, California, en route to Marine Corps Air Station El Toro, east of Orange County Airport. First Lt. James R. "Rick" Phillips was the pilot in command and 1st Lt. Christopher E. Schiess, in the rear seat, was RIO (Radar Intercept Officer). "At the time, Rick was 28 years old. He had previously been enlisted and then went back through OCS (Officer's Candidate School). Rick got a commission and went through flight training. We were flying by

Close-up of the F-4B's left wing with the main gear still retracted. The tip was torn away when the two aircraft collided.
(via Christopher E. Schiess)

ourselves that day because the other plane had been down for maintenance troubles. They went back down (to MCAS El Toro) the day before," Schiess said. The F-4B was returning to El Toro, its home base after an overnight cross-country flight to McChord Air Force Base, Washington. On the way back to El Toro, the Phantom made a stop for fuel at Mountain Home Air Force Base, Idaho. While the aircraft was on the ground, the crew had the radio inspected because it had failed on the landing approach. Maintenance personnel at Mountain Home replaced a fuse in the radio. However, they could not fix a small oxygen leak that they discovered. In addition, the transponder was inoperative. The Phantom's crew decided to file a VFR (Visual Flight Rules) flight plan to Naval Auxiliary Air Station (NAAS) Fallon, Nevada, to see if repairs could be made on the oxygen system. During the flight to Fallon, the oxygen leak increased. Ground personnel at NAAS Fallon were unable to make repairs. Phillips phoned his duty officer at El Toro for instructions and was advised to proceed home at low altitude.

After refueling, the crew filed a VFR flight plan to El Toro. Takeoff was delayed from Fallon because El Toro was holding an airshow, so 151458 did not depart until 5:16 p.m. The Phantom's route was to fly direct to Fresno, then follow Airway J-65 to Bakersfield, next to J-5 taking the jet fighter to Los Angeles, then transitioning Los Angeles' airspace direct to El Toro.

Once airborne, 151458 climbed to 15,500 feet to clear the eastern Sierra Mountains. On the western side of the range, 151458 descended to 5,500 feet and followed J-65 to Bakersfield. At this point the Phantom's crew elected to devi-

The F-4B Phantom II that collided with Air West N9345 was similar to the Air Force examples shown here patroling the Air Defense Identification Zone off the California coast.
(Nicholas A. Veronico)

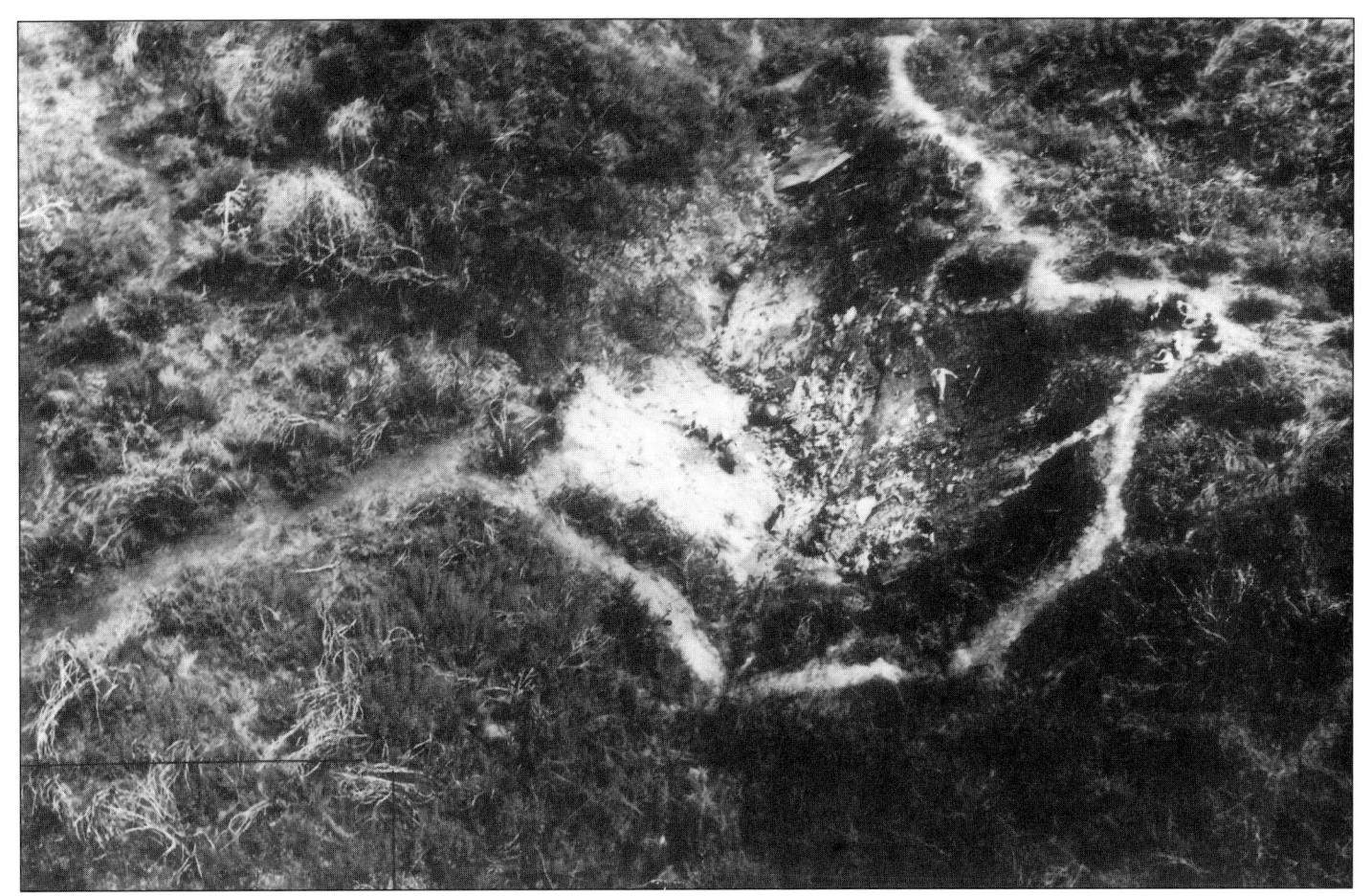

Investigators pour over the F-4B crash site. One engine and a wing panel are visible in the upper center. (NTSB)

ate from the flight plan. "We had filed to go a route that would have taken us right over Los Angeles," Schiess said. "And then we were talking back and forth and thought, 'Oh, geez, it's so crowded. Why don't we cut west of Los Angeles.' That's what took us to the east about 30 to 40 miles. (Looking back today,) it would have been a lot safer if we had flown right over the top of Los Angeles at 15,000 feet instead of being out where we were that put us in line with a SID (standard instrument departure)." They flew east of the planned course over Palmdale. Because of decreasing visibility at the lower altitudes and the fact that the San Gabriel Mountains were looming ahead, 151458 climbed once again to 15,500 feet. This was accomplished using maximum engine power without afterburner and took less then 2 minutes. Shortly after leveling off, and with the DME (Distance Measuring Equipment) indicating 50 miles to El Toro, Phillips executed a 360-degree aileron roll. This maneuver could be considered a questionable thing to do approaching the crowded skies of Southern California, and much was made of it in the press, although it had little, if any bearing on the event that was to follow.

While the Phantom was executing its climb to 15,150 feet, Air West Flight 706 was sprinting upwards at nearly 1,500 feet per minute en route to the Daggett VOR, its first checkpoint.

Both aircraft were closing very quickly. At one minute to impact, the two were nearly 10 miles apart. Thirty seconds to impact, 5.3 miles apart. If visible to each other, they would have appeared nearly stationary. In fact, the two aircraft were closing the distance at better then 1,050 feet per second, the speed of a .22-caliber bullet. Fifteen seconds before impact they were 2.6 miles apart. The smoky trail of the F-4B might have been visible at this point. Ten seconds prior, 1.8 miles apart, and five seconds before, less then one mile to impact.

Lt. Schiess Describes The Next Fateful Seconds

"It had probably been just a matter of minutes, like one to three, that we had been at that altitude (15,500 feet). We had been doing a low-level flight, 500 to 1,000 feet AGL (Above Ground Level), and then we just popped up. We probably went from 4,000 to 15,000 feet in a minute or two to get up top before we started our descent into El Toro. We got to the very top, did an aileron roll, and then that's when I got on the radar (and started) ground mapping," Schiess said.

In mapping mode, the RIO's head is down in the cockpit looking at the screen — only seeing what the radar is focused on. Schiess said, "You see the scope and the image of the ground and where it rises, depending on where you have the antenna pointed. This was equipment that is 30 years old now, and it did not provide high-definition or high-resolution images like they have today. That antenna had a 3-degree beam. At a couple miles, if you are at 15,000 feet, that might paint something up to 18,000 and down to 12,000 feet. You had to put the antenna where you want to see.

"I had my head down in the cockpit for 45 seconds to a minute, as if you were reading something, and looked up to

National Transportation Safety Board photo of the DC-9 impact point. Note the investigators on the 60-degree slopes of the canyon. The tail section is at the bottom of the gulley on its left side. (NTSB)

give my eyes a rest. I just happened to look up — I saw it and yelled instantly. I think the pilot probably saw it at the same time because his reaction was too quick for it to be a response to my yell. I shouted, 'Watch it Rick!' and he max'd left aileron deflection — slammed it against the left hand stop. We rolled extremely hard and nose low a little bit. I was watching and, for a mili-second, I did not know if we had missed or not.

"Then we hit. The back part of our airplane collided with the underneath of the DC-9 and it severed our jet in half.

"The NTSB report says that we were in a 28-degree roll when we hit, and I think that report is wrong. I know I lost sight of the airplane, and I know I could not have possibly lost sight of them if we were only at 28 degrees of roll. We had to be over 90. It is sort of amazing to me that science comes out wrong on that because they do have a lot of accuracy (in the report). But I would go to my grave saying they missed that one because I vividly remember losing sight of the airplane and wondering for a mili-second what was going to happen.

"There was absolutely no doubt there was a collision. Our airplane started tumbling, almost somersaulting. I said to myself, 'This airplane isn't going to fly!' Because I was getting bounced around a lot — my head was slamming against the canopy, I saw not only my warning lights (I had eight to 12), I saw a bunch of warning lights in the front. I thought to myself, 'Oh my God, Rick's already out.' That's when I pulled my ejection ring.

"Our airplane was going down in a fireball. I think your instinct is to just get out before weighing the consequences of what had just occurred. Your instinct is to get out of that airplane that you know is on fire and is destroyed.

"I reasoned incorrectly with respect to the fact that Rick would be out because the sequence is the back canopy goes, then the back seat goes, then the front canopy goes, then the front seat. The only way he would have been out was if my seat had malfunctioned. The best knowledge is that Rick did not try to get out before I got out. The reason he did not get out is one of three possibilities: he didn't try to get out and rode it in; he was knocked unconscious by the collision — but that's very unlikely in my opinion although I never attempted to talk to him after the collision because things happened so quickly, and the third, and most likely reason in my opinion, is that his seat malfunctioned. When my canopy went off, it caused a pressure differential and he was stuck in the aircraft. After that accident they put charges on the canopies so that, if in fact that was the case, it would not happen again.

"The seats had a barometer that was set to separate it from the pilot at a certain altitude. It should deploy at about 15,000 feet. I remember hearing the canopy blow off, going up and getting caught in the wind stream. I tucked my arms in as hard as I could so they would not get blown off, and then I opened my eyes, looked down, and saw the ground. I had enough time to realize that I still had my seat with me. It had not separated automatically and I got to thinking, 'Ok, I'd better try to do this manually.' At that time it separated automatically.

"I'm coming down in my chute and I've got airplane parts all around me — part of the vertical stabilizer fell — I don't know how close, but it wasn't very far from me. Then I looked down in time to see our airplane that was totally on fire. The first thing I started looking for was another chute. Then I watched our airplane until it hit the ground. At that point I started looking for the DC-9. I saw it and I watched it for 10 or 20 seconds. It looked like a leaf falling from a tree, just oscillating about 20 degrees on the vertical yaw axis. Nose level, going left and right, all the way down at a 60 to 70 percent rate of descent. Everything looked fine but it did not have any power. Everything I could see from looking above it was totally intact."

Having had most, if not all of its control cables and electrical systems from the cockpit to the rest of the aircraft severed, the DC-9 was doomed. Flight 706 entered a flat spin and impacted into the bottom of a steep canyon leading off of Mt. Bliss, about three-quarters of a mile from where the F-4B crashed. There were no survivors on the DC-9.

The Marine Corps F-4B Phantom II crashed approximately three-quarters of a mile from the DC-9. (via Christopher E. Schiess)

"I landed three or four miles from the DC-9 and one or two from our airplane," Schiess said. "I tried to have some control over my descent because I was coming down right on top of these huge powerlines. I thought, 'What an irony. I survived this collision and now I'm going to get fried by these powerlines.' I started yanking on the chute cords trying to change my angle of descent. It seemed to have no effect, but then I drifted off of them a little bit. I landed 50, maybe 75 yards below this Forest Service road. I just sat there and contemplated what had happened for a while and then a truck came by and picked me up."

The NTSB's Findings

The NTSB lists the probable cause of this accident as the failure of both crews to see and avoid each other, but recognizes that they only had marginal capability to detect, assess, and avoid the collision. Other factors include a very high closure rate, the combination of IFR and VFR traffic in an area where the limitation of the ATC system precludes effective separation of such traffic, and the failure of the crew of 151458 to request radar advisory service, particularly considering the fact that they had an inoperable transponder. If 151458 had requested the radar advisory there is a very good chance that the accident never would have occurred.

Schiess debates those points and said, "There was nothing wrong with the airplane that had any impact whatsoever with respect to the accident. That was overplayed by the news media. The transponder was not working, but I don't think it would have mattered. I don't believe they had altitude on transponders 22 years ago so that would be total conjecture. To conjecture that the transponder could have prevented the collision means that somebody on the ground would have noticed the signal and then alerted the other aircraft to our presence. Maybe it would have happened and maybe it wouldn't. It would have been extremely lucky for me to have picked up that airplane on radar because at the time of the accident I was trying to do some ground mapping with it. So I did not have the antenna aimed at where that aircraft was coming from. That probably had nothing to do with it. The oxygen system squawks were not a factor because the cabin was pressurized."

Follow-up And The Crash Site Today

Schiess stayed with the same squadron and continued his training until "combat qualified." This takes about a year while completing a syllabus in the aircraft type assigned. "I

Shortly after the crash, an attempt was made to remove the wreckage. A helicopter reportedly picked up the tail section and began to fly it out to be scrapped. The tail section, loaded in a sling, began to lift and the helicopter crew was forced to drop it about one-half mile from its original resting point.
(Michael B. McComb)

was about halfway through the syllabus when the accident occurred," Schiess said. "Six months later when I became combat qualified, I went to a group billet. At that time, one-third to one-half of the crews were getting orders to go overseas for eventual duty in Vietnam. Six months earlier everybody got orders to go to Vietnam. Six months later, nobody did. Because there was all this litigation due to the crash, I was held back. A few months later it did not make any difference because nobody was going over there. It was all winding down."

After leaving the Marine Corps, Schiess went onto a successful career in the pharmaceutical industry. He now is married with three children.

Michael B. McComb located the wreckage of Air West Flight 706 in fall 1987 while scouting the area in a Cessna 150. He captured the tail section's coordinates in his Loran at 34 11'30" North 117 56'10" West.

McComb said, "The aerial search lasted a mere 45 minutes. However, I had many hours of research invested in the area's libraries reading through newspaper micro film.

"I obtained a USGS Azusa topographic map and then went to work. I located Mt. Bliss, a prominent geographic feature mentioned in the newspapers and official reports. Once the approximate location and elevation was determined, I set out from the Upland/Cable Airport. Flying northwesterly toward the heart of the rugged San Gabriel Mountains, I began to see the silhouette of Mt. Bliss through the late-afternoon smog.

"At first I made a series of orbits around the mountain, making sure I kept a safe distance between myself and the terrain. During several turns around the mountain, I failed to see any metal fragments on the brush- and tree-covered slopes. The sun was beginning to set and I decided to return to the airport, 15 miles away. It was shortly after I made the turn that I looked down one more time and saw the unmistakable view of a DC-9 tail section sitting upright among the tall trees.

"I quickly logged the location and began a descent toward the find. Because of the terrain, I had to plan every turn I made within the narrow canyons. The tail section was within a ravine on the eastern face of Mt. Bliss facing Fish Canyon. I made several passes taking pictures and searching for additional wreckage. During these observations and later examining the photographs, I noticed that the Air West livery has been erased by the elements."

Mt. Bliss is 3,720 feet tall and the tail section rests between 2,600 and 2,800 feet. Shortly after the accident, an attempt to remove the tail section reportedly was made. Using a helicopter, the tail was being flown out from the crash site when it gained lift and was dropped before the helicopter became unstable and crashed. Thus, there are now three sites to investigate — the impact points of the F-4B, the DC-9, and the tail section. Schiess describes the F-4B's crash site, "I saw the pictures of my aircraft and there wasn't anything very big left. Of course, half of it was gone before it hit the ground."

This is rough country, but most Wreck Chasers will find the trip worthwhile.

N9345's tail section at rest on the slopes of Mt. Bliss. Photo taken from a Cessna 172 in December 1987.   (Michael B. McComb)

WC

The Mystery of TWA Flight Three
Actress Carole Lombard Perishes In Wartime DC-3 Crash

By Donald B. McComb Jr.
and
Michael B. McComb

Above: *The tail section rests inverted on a ledge of Mt. Potosi. DC-3 NC1946 took actress Carole Lombard and 21 others to their deaths on Jan. 16, 1942.* **Below:** *Clark Gable, center, leaves his Las Vegas hotel as plans for the funeral of his wife Carole Lombard are completed. With Gable are Howard J. Mannix, left, general manager and vice president of Metro Goldwyn Mayer studios, and friend Al Menasco, right.* (Both UPI/Bettmann)

On Jan. 16, 1942, at 7:07 p.m. PST, a TWA DC-3, NC1946, took off from the Las Vegas Western Airlines Terminal with 19 passengers and a crew of three for what should have been a routine flight to Burbank, California. Twenty minutes later, the DC-3 was a mass of flaming wreckage on the steep slopes of Mt. Potosi, 35 miles southwest of Las Vegas. The passenger list included 15 Army Air Corps servicemen, the wife of an Army Air Corps pilot, and notably actress, Carole Lombard, her mother Bessie Peters, and Otto Winkler, Metro Goldwyn Mayer (MGM) press agent for Lombard. The crew consisted of Captain Wayne Williams, First Officer Morgan A. Gillette, and Hostess Alice Getz.

This accident had most of the makings of a 1940s Hollywood movie. It included one of the most famous actresses of the time, traveling on a government mission. She carried thousands of dollars worth of expensive jewelry. An internationally famous violinist was bumped from the aircraft at Albuquerque and later suspected of being a enemy agent and saboteur who caused the crash. An FBI investigation of the accident was conducted by no less than J. Edgar Hoover himself, with the results of the investigation being followed closely by the White House.

The story begins on the morning of Jan. 12, 1942. The United States had been at war with Japan and Germany for just over a month. In Hollywood, a call for action went out. Clark Gable was picked to head the Hollywood Victory Committee. Wanting to do her part to help with the war effort, Gable's wife, Carole Lombard, volunteered to go on a fund-raising drive to sell war bonds for the Treasury Department. Lombard's goal was to raise $500,000 on a round-trip rail journey from Los Angeles to Indianapolis, Indiana, in her home state. Departure was set for the morning of the Jan. 12 and she was to return to Los Angeles' Union Station eight days later.

Accompanied by her mother and press agent Winkler, the trio departed Union Station as scheduled. Making several fund-raising stops along the way, they arrived in Indianapolis on Jan. 15. In a one-day flurry of appearances, Lombard gave speeches, signed autographs, and managed to exceed the Treasury Department's goal considerably. She raised a total of $2,107,513 for the war effort. Now at the end of what had been a very long day, Lombard was tired. She was not looking forward to the long trip home with the additional fund-raising stops that had been added. She wanted to fly home. Her mother, who had never flown before, was against this idea. Winkler also attempted to talk Lombard out of flying by saying that she could get more rest on the train. Lombard was adamant, "I'll sleep on the plane, and we'll be home tomorrow." At around 1 a.m., she finally decided to let the matter rest on the flip of a coin: Heads for the train, tails the plane. The coin landed tails up.

After calling the airport, they discovered three cancellations on TWA Flight 3, inbound from New York, and due to depart Indianapolis at 4:00 a.m. Several stops later, the flight would arrive at L.A.'s Burbank Airport 17 hours after leaving Indianapolis. When they arrived at the airport another debate took place between mother and daughter about flying home. Lombard's mother, Mrs. Peters, was a strong believer in the occult, mainly astrology and numerology. Consulting her astrology charts, she pointed out that it was now Jan. 16, and that the number "16" was a sign of impending accident or death. In addition, there were, according to Peters, "too many threes" involved in this trip. They were a party of three, they

Above: *The sheer rock face of Mt. Potosi is almost 90 degrees at the point of impact.* **Below:** *Soldiers prepare to lower the bodies of those killed in the crash.* (Both UPI/Bettmann)

were booked on Flight 3, a DC-3, and Lombard was 33 years old. Peters believed that "the number three is an unlucky number." In addition, Peters' personal astrologer had warned her to stay off airplanes in 1942.

At 3:16 a.m., the TWA DC-3 taxied up to the terminal at Indianapolis Airport. The DC-3-362, serial number 3295,

NC1946 was delivered new to TWA in March 1941. After refueling, the passengers boarded and Flight 3 departed westbound into the predawn darkness.

Flight 3 next stopped at Wichita, Kansas, where several passengers deplaned, and a few others boarded. Among them was Joseph Szigeti, an internationally recognized concert violinist. Szigeti had fled Hungary to escape the war in Europe. He was heading to California to take part in the Irving Berlin film "Holiday Inn," in which Bing Crosby introduced the famous song "White Christmas."

After departing Wichita, Flight 3 headed for Albuquerque, New Mexico, the last scheduled stop before Burbank. Prior to the flight's arrival, TWA agents in Albuquerque received orders from the War Department stating that 12 members of the Army Air Corps with priority tickets had to be given seats on Flight 3. According to records, five of the westbound passengers would have to be bumped from the flight. When Flight 3 landed, TWA Station Agent Ed Knudson informed the passengers of the situation. This was not well received. The Lombard party was asked to give up their seats. Winkler and Peters agreed, however Lombard, nearly home, was not about to be stopped. She told the TWA agent that she had been on a government mission, and that he would have to bump someone else. Because Lombard retained her seat, the other two members of her party were not bumped, and decided to continue on to California.

The only other civilian to retain a seat on Flight 3 was Mrs. Lois Hamilton, wife of an Air Corps pilot. Among the passengers deplaned at Albuquerque was violinist Szigeti. Also at Albuquerque, a crew change took place, with Williams, Gillette and Getz boarding. Before the flight could leave Albuquerque, one other problem had to be solved.

The Army Air Corps pilots and enlisted men had a lot of heavy luggage, duffel bags, parachutes, etc. This made the DC-3 heavier than allowed by federal regulations. In order for the flight to be legal, TWA agent Knudson made some "adjustments" to the figures in the paperwork. According to government records, Knudson counted four of the passengers as weighing 150 pounds each. In addition, he asked Lombard, Peters and Hamilton their actual weight. Using these new figures, the aircraft was now "legal" for flight, but in actuality still weighed more then the CAA (Civil Aeronautics Administration) rules allowed.

At the time Captain Williams took command of Flight 3 on the afternoon of Jan. 16, 1942, he had over 12,000 hours of flight time. However, he had less then a perfect record as a pilot. In June 1929 he went to work for American Airways (predecessor to American Airlines). Late in 1931 Williams applied for a job with TWA, but was turned down because of word in the industry that he took unnecessary risks with company equipment and cargo. Nevertheless, shortly after the interview, TWA hired Williams as a night mail pilot. Within two years, he was fired, after being charged with damaging equipment, carelessness and insubordination.

Williams was terminated because the official at TWA who had originally hired him changed his mind. Citing constant violations of air safety rules, the fall 1933 memo concluded, "I do not care to be responsible for his flying with passengers."

Less than 60 days later, Captain Williams was rehired after the National Labor Relations Board ruled that his rights had been violated and that he should be reinstated, but as an airmail pilot only. During the next nine years Captain

Author Michael B. McComb poses with one of the DC-3's engines near the point of impact. (Michael B. McComb)

Williams' record showed no major complaint or disregard for the rules.

Departing Albuquerque

Now more then three hours behind schedule, Flight 3 finally departed. Because of strong headwinds that had been predicted for the route, Captain Williams decided to make two additional stops for fuel, at Winslow, Arizona, and Boulder City, Nevada.

Shortly after takeoff the crew of Flight 3 decided that they could save time by changing their flight plan. Captain Williams transmitted a request to TWA flight operations in Burbank. After recalculating fuel consumption Williams decided not to stop at Winslow for fuel. However, because the flight was behind schedule, they would be arriving after dark at Boulder City Airport, which had no landing lights. Captain Williams requested permission to land at Las Vegas. These changes were granted at 4:38 p.m. PST.

At 6:36 p.m. PST, Flight 3 taxied up to Las Vegas' Western Airlines Terminal building, located at what is now Nellis Air Force Base — this was an era when most airlines had their own airports. For this reason, the TWA station manager drove over from the company's airport at Boulder City to meet the flight. While the passengers went inside to grab a sandwich and something to drink, the DC-3 was fueled with 225 gallons of gasoline and oil was added to the engines. A few minutes before 7:00 p.m., the passengers boarded the DC-3 for the last leg of the trip.

At 7:07 p.m., NC1946 departed Las Vegas and turned towards the southwest. Captain Williams radioed back to the airport that they had "cleared the field" and were proceeding on course.

What happened next at 7:23 p.m. is known. How it happened is not as clear. Climbing through 8,000 feet at over 150 miles per hour, the left wing of the DC-3 clipped a rocky ledge, cartwheeling the aircraft into the face of a cliff. With a thunderous explosion, hundreds of gallons of aviation gasoline detonated, scattering wreckage, bodies and luggage down the face of the cliff. Most of the debris came to rest at the bottom of a ravine. Had Captain Williams been 550 feet higher he would have missed the top of Mt. Potosi with 15 feet to spare.

Accident Investigation

How did this happen? To this day no one really knows. Because this accident took the life of Carole Lombard, a personal friend of President Roosevelt, and because this accident occurred during wartime under less-than-clear circumstances, federal investigations on several levels were begun. The Civil Aeronautics Board (CAB) launched a conventional aircraft investigation into the crash. The Federal Bureau of Investigation (FBI) looked into the possible criminal aspects of the accident to determine if the aircraft had been sabotaged. This investigation was headed up by none other than FBI Director J. Edger Hoover. In addition, a Congressional probe reviewed the circumstances surrounding the accident.

Top: *One of the landing gear with exposed oleo imbedded in the rocks of Mt. Potosi.* (Michael B. McComb)

One of the first things the FBI did was examine the aircraft wreckage for evidence of bomb damage. None was found. Another area scrutinized was the background of the passengers who had departed the aircraft at prior stops. One person who came under the FBI scrutiny was violinist Joseph Szigeti. Having come to the United States from Hungary, and having left the aircraft in New Mexico, immediately he was suspected of being a saboteur. For months the FBI probed into his past. Long after the investigation was closed, the FBI continued to read his mail. In the end, Szigeti was cleared.

Above: *Aerial view of Mt. Potosi.* (Michael B. McComb)

Author Michael B. McComb prepares to head to the DC-3's crash site. (Michael B. McComb)

CAB investigators noted errors in the flight plan filed by the co-pilot. This document still had all of the altitude settings and coordinates for the take off from Boulder City. The altitude filed was for a cruising altitude of 8,000 feet. This seems low even for a flight out of Boulder City, as the mountains to the southwest reach upwards of 6,000 feet. However, with the route nearly 20 miles further north, the flight path crosses more into the heart of the Charleston range where there are several peaks well over 8,000 feet in height.

Another contributing factor may have been the fact that since the bombing of Pearl Harbor, the beacon system had been blacked out throughout the western United States. This could have caused confusion in the cockpit as to their location in relation to the mountains. It has been suggested that possibly the crew was adjusting the throttles, in which case the cockpit lights would have been up to make it easier to read the instruments. Some of the pilots who knew Captain Williams had speculated that no sooner had the DC-3 left the ground, that he was back in the cabin visiting with Lombard, leaving an inexperienced co-pilot flying the aircraft. This scenario does not make sense. First, Captain Williams had more then enough time to visit with Lombard between Albuquerque and Las Vegas while it was daylight. Second, as far as the co-pilot being inexperienced, nothing could be farther from the truth. Granted, he did not have the 12,000-plus hours that Captain Williams possessed, but he did have over 3,000 hours total time; 755 had been accumulated during his year and one-half with TWA. He also was rated as a co-pilot on the Boeing 307 Stratoliner, which was a state-of-the-art aircraft in 1942.

The bottom line on this accident is that we never will really know what happened to cause an highly experience flight crew to fly into a mountain. The official verdict in the CAB report stated: "Upon the basis of the forgoing finding, and of the entire record available at this time, we find that the probable cause of the accident to aircraft NC1946 on Jan. 16, 1942, was the failure of the captain after departure from Las Vegas to follow the proper course by making use of the navigation facilities available to him."

During the week following the accident, wreckage was examined and the bodies of the passengers and crew were removed. A special effort was made to find several items of jewelry that Carole Lombard had with her. Clark Gable, who had flown up to Las Vegas hours after the accident in a chartered Western DC-3, was particularly interested in recovering two diamond and ruby clips that he had given Lombard for Christmas a few weeks earlier. On Jan. 20, a soldier approached a Gable associate and offered to sell a piece of the jewelry that he had found. This item was confiscated and the incident was reported to the soldier's commanding officer. This was believed to be the only piece of jewelry recovered by the search teams after the accident.

The Crash Site Today

More than half a century has passed since that dark, cold January night claimed those aboard Flight 3. Time has eroded the crash site and the memories of those who were involved in the aftermath of the accident. Most of the major figures have been dead for years. Some of the FBI files are sealed to this day, and those that are available have been heavily censored. In researching this accident, extensive use was made of newspaper accounts, government accident reports, and historical records from Clark County, Nevada.

Locating this site was very difficult. In newspaper articles Mt. Potosi goes by several different names including Double-Up Peak, Table Spring Mountain, and Olcott Mountain, to name a few. After viewing old pictures of the site, the location was pinpointd by taking aerial photos of Mt. Potosi and comparing them. No wreckage could be seen from the air.

On Oct. 19, 1992, an expedition was mounted. After a two and one-half hour hike, the lower crest of Mt. Potosi was reached.

From the elevation contour of 7,700 feet, I (Michael B. McComb) proceeded to hike along the base of the fossil-encrusted cliff that the DC-3 had impacted. The hiking was extremely difficult. There were very few level spots; the ground was crumbly and the incline reached in excess of 50 degrees. Numerous ponderosa pine trees dot the side of the mountain, some reaching upwards of 75 feet. This area of the Mt. Charlston range is considered the most rugged in Nevada, if not the western United States.

Michael B. McComb digging at the impact point. Note the steepness of the terrain. (Michael B. McComb)

After an additional hour of hiking, I clearly could make out the distinctive shape of a Wright Cyclone engine, its cylinders and pistons frozen by the impact. Other bits of airframe could be seen scattered around. In a scrub oak brush lay one of the main gear, its oleo struts still gleaming. Not far away was a prop dome, pointing towards the sky it would never fly in again. Down slope was the flap from the right wing. There were little items too, like a crushed Stanley thermos. Stamped on the base was the company's motto "It will not break." However in this case...it did. Not far away was a piece of plastic drinking cup with the letters "TWA" on the base.

Large pieces of fuselage skin can still be found at the crash site more than 50 years after the accident.

Above and Right: *Right: Engine and firewall components litter the area.* (All photos this page by Michael B. McComb)

In early September 1993, I made another trip to the site. This time I gained access to a utility road that made possible a more efficient route to the location. From the trailhead where I parked, the crash site is a short but very hard one-mile hike up the very steep slope of Mt. Potosi. The first thing I accomplished was to locate the point of impact on the cliff. Next I found where the aircraft cabin came to rest and burned. This was accomplished by finding cabin materials, such as silverware, dishes, and personal effects including coins and jewelry.

The tail section with the words "Sky Club" painted on it was no longer there — or was it? While examining some of the wreckage I found a piece of airframe about 8 inches by 8 inches. The piece had a red stripe with black borders, consistent with the size of the "Sky Club" letters on the tail. The tail section, apparently along with most other larger fragments, had been broken up to avoid being reported as a new crash.

Also among the wreckage I found some shovels and sifting screens. These items showed signs of having been there at least a few years, no doubt left at the site by somebody looking for trinkets, or perhaps Carole Lombard's lost jewels.

At the point where I judged the main portion of the fuselage had come to rest, I laid out a 10-foot by 15-foot area to be excavated. The techniques I utilized are similar to those used at any archaeological excavation. The soil is placed in a screen and shaken and sifted. Among the odd bits and pieces located using this method were jewelry items such as a gold ring, pieces of a gold hairclip, and a small piece of a gold watchband. In addition, a large emerald cut synthetic ruby with a weight of 14.5 carats was found. Also found was a small star sapphire. Had the ruby been a natural one, its value could been in excess of $50,000. Amid the bits of debris were several Army Air Corps "wings and propeller" collar insignia, an rpm gauge face, and a Sperry Gyro Horizon instrument face. A medal I found gave me cold chills. On the obverse it said, "St. Joseph protect us." On the reverse it said, "My guardian angel be my guide." It should be noted that the excavation of the site was not for the purpose of treasure hunting but to locate historical material and aviation related material from that time period.

A few weeks before her fateful departure for Indianapolis, Carole Lombard had just finished making a movie with Jack Benny titled "To Be or Not To Be". When the movie was released in February 1942, the producers edited six words from the original version. Removed from the film was a scene in which Carole Lombard asked Jack Benny, "What could happen on a plane?"

WC

Ill-Fated Airliners

I ll-Fated Airliners presents a selection of aircraft prior to their last flight. Along with these rare photographs, a brief description details each plane's demise. Types shown range from the 1930s vintage prototype Boeing Model 307 Stratoliner to today's mamoth 400-passenger 747-400s.

Boeing 747-258F 4X-AXG (msn 21737) ▪ El Al ▪ Amsterdam ▪ Oct. 4, 1992

El Al 747-258F 4X-AXG (msn 21737) was delivered on March 19, 1979. Configured as a freighter, the plane flew with a crew of four. On the evening of Oct. 4, 1992, 4X-AXG took off from Amsterdam and immediately both starboard engines lost power. The crew attempted to return to the airport but crashed 10 miles short. The fully loaded 747 struck an apartment building in Bujlmermeer. All four crew members were killed along with 80 apartment residents. 4X-AXG is seen on approach to Los Angeles International Airport (LAX) a few short months before the accident. *(Stephen L. Griffin)*

Boeing 747-409 B-165 (msn 24313) ▪ China Airlines ▪ Hong Kong ▪ Nov. 4, 1993

This China Airlines Boeing 747-409 overran the runway at Hong Kong and came to rest in the harbor. Its tail was an obstruction on the approach end of the runway and officials decided to dynamite the vertical stabilizer off until the aircraft could be moved. The aircraft was a total write-off. *(Stephen L. Griffin)*

Lockheed L-1011-1 N726DA (msn 1163) ▪ Delta ▪ DFW Airport, Texas ▪ Aug. 2, 1985

Delta Flight 191 from Fort Lauderdale, Florida, to Los Angeles, California, was on approach to Dallas/Ft. Worth, its only stop, when struck by the meteorlogic phenomenon windshear. The L-1011 entered the windshear area and was buffeted about by winds from all directions. The plane touched down 6,000 feet short and 360 feet left of the runway centerline in a field. The jet became airborne again and struck a car, killing the driver as it crossed a highway. Eight crew and 137 passengers were killed while 30 others sitting mainly in the tail area survived. *(ATP/Airliners America)*
(For further information, see NTSB Report Number AAR-86-05.)

Ill-Fated Airliners

Boeing 707-331C N15712 (msn 20068) ▪ TWA ▪ Sept. 13, 1972

TWA's N15712 made its first flight on June 18, 1969. Delivered to the carrier two weeks later, on July 2, its career ended in the salt water of San Francisco Bay after an aborted take off on Sept. 13, 1972. *(ATP/Airliners America)*
(For further information, see NTSB Report Number AAR-73-04.)

Boeing 727-22C N7434U (msn 19891) ▪ United ▪ Los Angeles, Calif. ▪ Jan. 18, 1969

Photographed in early September 1968 on Boeing's Renton, Washington, flightline, N7434U was delivered new to United Airlines on Sept. 20, 1968. Less than four months later, on Jan. 18, 1969, the plane crashed shortly after take off from Los Angeles International Airport (LAX). The plane fell into the Pacific Ocean less than four minutes after becoming airborne. *(ATP/Airliners America)*
(For further information, see NTSB Report Number AAR-70-06.)

Ill-Fated Airliners

Prototype Stratoliner Lost During Demo Flight
Boeing 307 NS19901 (msn 1994) Boeing Prototype ▪ March 18, 1939

The first pressurized airliner, Boeing's model 307 Stratoliner, revolutionized commercial aviation. Pressurization allowed airliners to cruise at 20,000 feet — above harsh weather conditions, while providing a cabin atmosphere near sea level. Boeing's new airliner would carry 33 passengers in a day configuration or 25 in Pullman-style berths.

The Stratoliner first flew on Dec. 31, 1938, in the capable hands of famed Boeing Test Pilot Eddie Allen. While additional aircraft rolled down the assembly line, Allen and another Boeing pilot, Julius Barr, wrung out the new airliner. During a demonstration flight for representatives of Royal Dutch Airlines on March 18, 1939, the prototype, NX19901, was lost. Eyewitnesses said the 20-ton, $500,000 airliner "appeared suddenly out of the clouds and that the sound of its motors died down momentarily. The motors then seemed to speed up and the plane began a long, crazy spin earthward." Government investigators attributed the crash to an "abrupt pullout from a dive following recovery from an inadvertent spin." Killed in the crash were test pilots Barr, Earl A. Ferguson and William Doyle, Boeing Aerodynamicist Ralph L. Cram, Boeing Sales Representative Ben Pearson, Boeing foreman Harry C. West Jr., Transcontinental and Western Air Chief Test Pilot Harlan Hull, Netherlands State Airworthiness Board Engineer A.G. von Baumhauer, and Royal Dutch Airlines Assistant General Manager Peter Guilonard.

Top: *The prototype Boeing Model 307 Stratoliner lies in canyon near Alder, Washington, after a stall and failed recovery attempt. Representatives of Royal Dutch Airlines and a number of Boeing's top engineers and aerodynamicists were killed in the crash. (Gordon S. Williams via John M. Campbell Collection)* **Above:** *Prototype Stratoliner NX19901 in flight. (Gordon S. Williams via Museum of Flight)*

Ill-Fated Airliners

Hawker Siddeley (HS121) Trident 3B G-AWZT (msn 2320) ▪ British Airways ▪ Sept. 10, 1976

Trident 3B G-AWZT flew its maiden flight on May 26, 1972, and was delivered to British European Airways June 6. Merged to become British Airways on April 1, 1974, the tri-jet met its end two years later in the skies over Yugoslavia. Flight 476 took the Trident from Heathrow to Istanbul, Turkey, with nine crewmembers and 54 passengers. Air Traffic Control procedures broke down at the Zagreb Area Control Center and Inex Adria Airways Douglas DC-9-32 YU-AJR (msn 47649), on a charter to Cologne, West Germany, was approved for the same flight level as the Trident. Both aircraft met head-on at 33,000 feet, the DC-9's left wing slicing through the Trident's flight deck. Both airliners fell to earth outside of Vrobec Village, near Zagreb. The DC-9's five crew members and 108 passengers brought the death toll to 176.

(ATP/Airliners America)

Boeing 707-3B5C HL-7406 (msn 20522) ▪ Korean Air Lines ▪ Nov. 29, 1987

Delivered new to Korean Air Lines on Aug. 6, 1971, HL-7406 served the airline until its crash on Nov. 29, 1987. Operating as Flight 858 from Baghdad, Iraq, to Seoul, South Korea, it disappeared while crossing the Andaman Sea near Burma. The plane last had stopped in Abu Dhabi, United Arab Emirates, and was to stop again at Bangkok, Thailand, before continuing on to its final destination. Everyone on board, 114 passengers and 11 crew, perished. An eyewitness described the aircraft exploding and falling into the sea. Two bombing suspects were being sought when they both attempted suicide. One survived, and eventually recovered, later admitting that the pair had enplaned at Baghdad and smuggled a bomb on board disguised as a portable radio. The suspects left the bomb in an overhead bin and deplaned at Abu Dhabi.

(ATP/Airliners America)

Ill-Fated Airliners

Douglas DC-10-10 N110AA (msn 46510) ▪ American ▪ Chicago, Ill. ▪ May 25, 1979

    A failure to thoroughly check engine attach points on Douglas DC-10-10 N110AA (msn 46510) resulted in the loss of Number One on take off at Chicago O'Hare International Airport on May 25, 1979. When the powerplant and engine pylon separated from the wing, it took with it a number of hydraulic lines and the retraction cylinders for the inboard leading edge slats. Air pressure forced the port wing slats to retract, causing an asymmetrical lift condition which resulted in a roll to the left. The crew was unable to recover from the maneuver and the DC-10-10 crashed into a field, taking 258 passengers, a crew of 13, plus two on the ground to their deaths. *(ATP/Airliners America)*
(For further information, see NTSB Report Number AAR-79-17.)

Boeing 747-121 N739PA (msn 19646) ▪ Pan Am ▪ Lockerbie, Scotland ▪ Dec. 21, 1988

    N739PA was delivered to Pan Am on Feb. 15, 1970, and christened *Clipper Morning Light*. Ten years later the plane was renamed *Clipper Maid of the Seas*. A side cargo door was added in September 1987. Flight 103, a Boeing 727, departed Frankfurt, West Germany, on Dec. 21, 1988, for New York City with an en route stop at London's Heathrow Airport. Upgraded to a 747, Flight 103 took on an additional 194 passengers and their baggage. Its takeoff delayed at Heathrow almost one-half hour, the plane finally began its journey westward. En route to New York, the plane exploded 40 minutes into the flight over Lockerbie, Scotland. A bomb had been placed in a suitcase and detonated in the left forward cargo compartment. Had *Clipper Maid of the Seas* not been delayed, it would have plunged into the sea leaving investigators unable to solve the mystery of the crash. Searchers found that the bomb had been planted in luggage that was transported by an Air Malta flight to Frankfurt, then transferred to Flight 103 at Frankfurt. The bags were not re-screened at Heathrow prior to loading. Libyan terrorists were blamed for the disaster. *(ATP/Airliners America)*

Ill-Fated Airliners

Convair 600 N94230 (msn 56) ▪ Texas International ▪ Mena, Ark. ▪ Sept. 27, 1973

This aircraft originally was delivered to American Airlines as a Convair 240 (msn 56) *Flagship Columbus* on April 23, 1948. Later renamed *Flagship Kitty Hawk,* the Convair twin was sold to Trans-Texas Airways (later Texas International) on Feb. 20, 1961. In May 1967, Rolls Royce Dart 542 turboprop powerplants were added, updating the aircraft to Convair 600 configuration. Operating as Flight 655 on the night of Sept. 27, 1973, N94230 had veered off course during its scheduled trip from Eldorado, Arkansas, to Pine Bluff, Arkansas. The Convair struck Black Fork Mountain at 2,025 feet, more than 600 feet below the summit of the peak. Eight passengers and three crew perished in the crash.
(For further information, see NTSB Report Number AAR-74-4.)

(Airliners Slides)

Boeing 737-293 N468AC (msn 20334) ▪ Air Cal ▪ Santa Ana, Calif. ▪ Feb. 17, 1981

Sold by Boeing to GATX Leasing Corp. and subsequently leased to Air California (Air Cal) on Sept. 15, 1970, Boeing 737-293 (msn 20334) was registered N468AC. The plane met its end while operating as Flight 336 on Feb. 17, 1981, when it skidded off the runway at Orange County, California's John Wayne Airport. Another Air Cal 737, Flight 391, was given takeoff clearance, then instructed to abort when halfway down the 5,700-foot runway. At the same time, Flight 336 was instructed to go around due to traffic on the runway. When the cockpit voice recorder's tapes were reviewed, they showed that 15 seconds elapsed from the time the controllers gave instructions aborting the landing to the sound of power being applied to the engines. The aircraft pancaked onto the runway and slid to a stop after catching fire and breaking its back. N468AC came to rest less than 700 feet from the airport's fire station. More than 36 of the 104 passengers and five crew on board Air Cal Flight 336 were hospitalized. There were no fatalities.

N468AC was pushed off to one side of the runway to make way for landing traffic. Company officials quickly painted out the registration number and titles, but could not hide the 737's tell-tale livery. Msn 20334 was scrapped on-site.
(For further information, see NTSB Report Number AAR-81-12.)

(Scott A. Thompson)

Ill-Fated Airliners

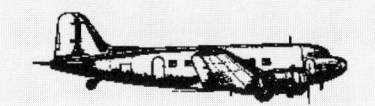

Douglas DC-9-15RC N565PC (msn 47240) ▪ Emery Worldwide Airlines ▪ Feb. 17, 1991

Continental was the first to operate msn 47240 when it took delivery of the plane, registered N8919, on July 20, 1968. The DC-9-15RC was McDonnell Douglas' "Rapid Change" configuration featuring palletized removable seats, converting the aircraft to or from a freighter. Hughes Airwest acquired the DC-9 and re-registered it N9351. Ownership passed to Republic Airlines when the two companies merged in January 1980. The plane was re-registered N565PC when bought by Purolator Courier Corp. in October 1984. It crashed on takeoff Feb. 17, 1991, at Cleveland, Ohio.

(Kevin Sullivan photos via Brian Sullivan)

(For further information, see NTSB Report Number AAR-91-09.)

Above: *N565PC came to rest inverted, killing both crew members. The fuselage split upon impact and the port side of the cockpit was crushed.*

Left: *A wrecker was brought in to clear the aircraft. Freight containers behind the wrecker's hoist still contain packages.*

Right: *Two cranes lift the aft portion of the DC-9-15RC prior to removal. Fireman standing near closed wheel well doors lends scale to the size of the wrecked DC-9.*

Ill-Fated Airliners

Douglas DC-10-10 TC-JAV (msn 46704) ▪ Turk Hava Yollari ▪ Senlis, France ▪ March 3, 1974

Turkish Airlines (Turk Hava Yollari) took delivery of DC-10-10 TC-JAV Ankara on Dec. 12, 1972. Less than two years later, the crash of TC-JAV with the loss of 346 lives, sparked a controversy over the design of the DC-10's cargo-door locking mechanism. TC-JAV had reached 11,000 feet when a cargo compartment door blew out, decompressing the cabin, which resulted in the structural failure of the cabin floor. Cables to the Number Two Powerplant, elevator and rudder, are assumed to have been disabled, causing a steep rate of descent. A recovery was being attempted before the DC-10 ran out of sky and impacted into a forest near the French town of Senlis, northeast of Paris.

(ATP/Airliners America)

Douglas DC-9-31 N1335U (msn 47393) ▪ Southern Airways ▪ New Hope, Ga. ▪ April 4, 1977

Double engine failure claimed Southern Airways Douglas DC-9-31 N1335U during a thunderstorm while en route to Atlanta, Georgia. Operating as Flight 242, the trip originated in Muscle Shoals, Alabama, with one stop at Huntsville, prior to its terminus of Atlanta. Tornado watches were in effect for the region with thunderstorms and 3-inch hail reported. After entering the storm cell near 16,000 feet, the crew reported that its Number One powerplant had failed and that the windshield was cracked by hail. Shortly thereafter, Number Two quit. The crew tried to restart the now silent fans, to no avail. Attempting to glide into Dobbins Air Force Base, the DC-9 finally ran out of airspace at New Hope, Georgia. Upon striking the ground, the aircraft broke apart and caught fire. A car was crushed and a number of people on the ground were struck and killed by debris. Nine residents of New Hope perished along with 63 of the 85 passengers and crew.
(For further information, see NTSB Report Number AAR-78-3.)

(ATP/Airliners America)

Ill-Fated Airliners

Lockheed 188A N9707C (msn 1099) ▪ Braniff ▪ Dawson Texas ▪ May 3, 1968

Departing Houston, Texas, on May 3, 1968, Braniff Flight 352 encountered a thunderstorm. Equipment for the flight was Lockheed 188A N9707C (msn 1099), delivered new to the carrier on Oct. 17, 1959. Given a weather briefing prior to takeoff, the crew attempted to fly through the storm despite attempted re-routing from Fort Worth's Air Traffic Control Center. Facing the storm head-on, the crew elected to reverse course 180 degrees and was granted permission to do so. A 180-degree turn in a severe storm was against Braniff policy because the turn's banking attitude can quickly turn into a stall when coupled with an up- or down-draft. Unfortunately for the 80 passengers and crew of five, this is exactly what happened. During its right-hand 180-degree turn, winds lifted the port wing, forcing the starboard wing down past 90 degrees. A stall resulted from this unusual attitude, pitching the nose down. During the recovery maneuver, the starboard wing was over-stressed, separating at the outer wing panel beyond the Number Four engine. The tail of the Electra separated, closely followed by the engines and control surfaces of the port wing. The aircraft came to rest in a field near Dawson, Texas. *(ATP/Airliners America)*
(For further information, see NTSB Report Number AAR-69-3 dated June 19, 1969.)

Boeing 707-437 VT-DMN (msn 18055) ▪ Air India ▪ Mont Blanc, France ▪ Jan. 24, 1966

Air India took delivery of VT-DMN on April 17, 1961, and christened the aircraft *Kanchenjunga*. Less than five years later, *Kanchenjunga* ended her flying career 200 feet below the sumit of Mont Blanc in the French Alps. Air India Flight 101 had departed Bombay and was making a planned stop at Geneva, Switzerland, en route to London. There was a crew of 11 plus 106 passengers, all of whom perished in the crash. The cause of the accident was determined to be a navigational error coupled with incorrectly set altimeters. *(ATP/Airliners America)*

Oakland Mid-Air: See And Be Seen
Wreckage Of One Crashes On Road, Second Lands Safely

By Nicholas A. Veronico

N79993 Fern M., *sistership to ill-fated Overseas National DC-4 (C-54D-10-DC) N79992 (msn 10832)* June S, *in front of California Eastern Airway's facility at Oakland Municipal Airport.* (William T. Larkins)

The early 1950s were heady days in commercial aviation. Thousands of experienced pilots were home from World War II seeking employment. The Korean conflict was on and hundreds of non-scheduled carriers flew cargo and passengers to feed the war effort. Two such carriers were California Eastern Airways Inc., and Overseas National Airways (ONA) Inc. Both airlines were incorporated in Delaware; Cal Eastern in 1946 and Overseas National in 1950. Each maintained a base at Oakland Municipal Airport, California. Cal Eastern's primary business was flying passengers and freight on "mid-Pac" routes as part of the Pacific Airlift for the Military Air Transport Service (MATS). Overseas National also flew government contracts.

On the morning of Nov. 17, 1951, each carrier sent a DC-4 aloft to give pilots their six-month instrument competency checks. Before lunch one aircraft would be destroyed and the second flyable, but heavily damaged. Three highly experienced Overseas National captains would lose their lives.

The Aircraft, Carriers, And Their Crews

Ill-fated Overseas National DC-4 (C-54D-10-DC) N79992 (msn 10832) was delivered as serial number 42-72727 to the U.S. Army Air Force on July 16, 1945. The plane was leased to ONA and registered in 1947. In 1949, Transocean Air Lines subcontracted to operate Overseas National's aircraft. This agreement was short-lived and the company was again flying its own equipment by April 1950. The plane had flown 5,257 hours since its last major overhaul. Although leased to ONA at the time of the crash, it was still owned by the Air Force. ONA's fleet name for N79992 was *June S*.

Crew for this flight included Captain Marion Harvey Click, 31, ONA's assistant chief pilot. Click was trained by the Army Air Force during World War II and went on to fly with a couple of carriers before joining the company on June 16, 1950. He had accumulated 7,871 hours, 5,028 in type. Captain Warren Cecil Gessner, 27, was trained by the U.S. Navy and flew with Alaska Airlines, Near East Transport and

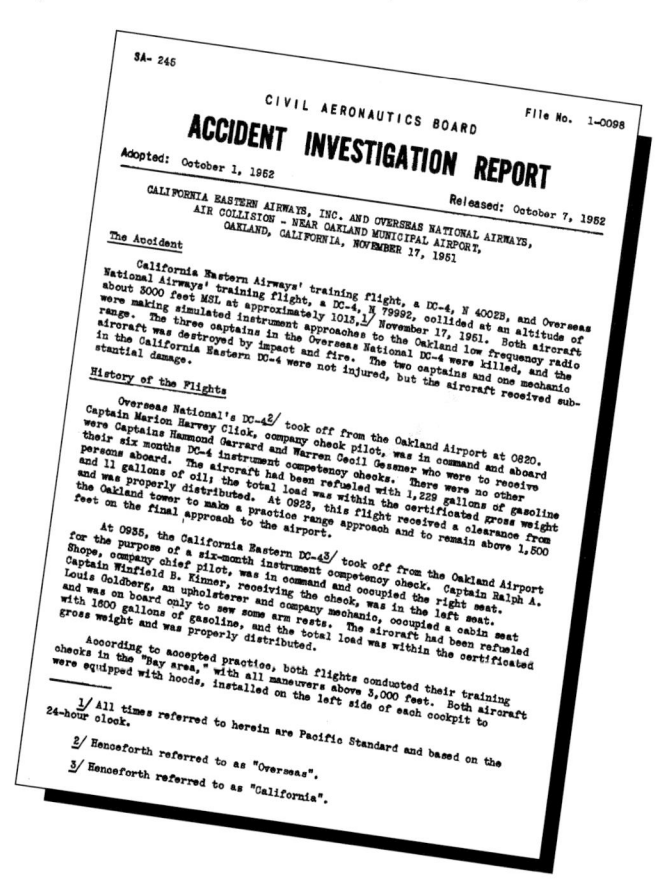

80

Douglas DC-4 in an early California Eastern Airways color scheme. Captain Ralph A. Shope was able to maintain control of his aircraft and land it successfully at San Francisco. (William T. Larkins)

Westair Transport prior to hiring on with ONA. His flying time was 3,590 hours, just over 1,900 in DC-4s. The third member of the crew was Captain Hammond Garrard, 32. His employment with the carrier was somewhat turbulent. He was hired on July 6, 1950; promoted to captain, July 17, 1950; but was then furloughed until Nov. 15, 1950. Like Gessner, he earned his wings of gold with the Navy, then flew for Capital Airlines and Cal Eastern. Also very experienced, he had amassed 5,879 hours, 3,100 in type, yet only 330 while flying on instruments. Click would serve as examiner while Garrard and Gessner performed their instrument proficiency checks.

Cal Eastern's DC-4 (C-54B-1-DC) N4002B (msn 10474), was delivered to the Army Air Force as serial number 42-72369 on Dec. 12, 1944. Just over one year later, on Dec. 20, 1945, Douglas repurchased the aircraft and converted it to civilian DC-4 standards, assigning the plane conversion Number 15. It returned to the skies flying for Western Air Lines on Aug. 8, 1946, as N88701, then was acquired by United Air Lines and christened *Mainliner Hoover Dam* one year later. Cal Eastern acquired the aircraft in 1950. Then title passed to TACA Airlines, Cal Eastern leasing the plane back on April 5, 1951. On the day of the mid-air collision, N4002B had flown 16,435 hours total, 6,484 hours since its last major overhaul.

Captain Ralph A. Shope, 40, Cal Eastern's chief pilot, began his flying career in 1934. He was an Army flying instructor, flew for TACA, and TWA-ATC (Air Transport Command) during World War II. Of his total 11,225 hours flight time, 4,733 were in type with 1,009 on instruments. On Nov. 17, 1951, he was flying as check pilot for Captain Winfield B. Kinner, 40, who began his flight training in 1928 and also was an instructor for Army cadets during the war. He later flew for ATC before joining a number of non-scheduled carriers. Kinner had accumulated 5,000 hours, 3,500 in type and 350 on instruments. While the two pilots performed their flying and checkride duties, Louis Goldberg, a Cal Eastern mechanic and upholsterer, flew in the cabin repairing the seats.

Nov. 17, 1951, A Beautiful Day

Being a Saturday, commercial traffic was light at Oakland. The weather was pleasant with thin broken clouds at 25,000 feet, visibility of seven miles, and a light breeze blowing from the south. ONA's N79992 took off at 8:20 a.m. Over an hour later, at 9:35, Cal Eastern's N4002B also departed from Oakland Municipal. Once at altitude, the pilot's side windows were covered with a "hood" to simulate instrument meteorological conditions (IMC). The check pilot observed from the right seat and the third captain on board N79992 acted as observer.

Both aircraft were flying at 3,000 feet, approaching the Oakland low-frequency radio range station at 10:13 a.m. — Overseas' N79992 on a magnetic heading of 124 degrees; Cal Eastern's N4002B's heading was 75 degrees. Observers on the ground saw both aircraft flying towards each other and converging without taking evasive action. ONA was slightly

The crash site today: Doolittle Drive facing northwest. N79993 came to rest approximately where the white arrow points.

81

This is how the crash site looked once the smoke had cleared. The tail (1), fuselage (2), wing and nose section (3) fell separately near Doolittle Drive. A car driven by Robert Leibel (4) struck the aircraft as it hit. Two other cars swerved into the parking lot at right as the wreckage fell toward them. Norman Barnes' car (5) was scorched, but he and his son escaped injury.
(Inset photo by Bob Hofford, main by San Francisco Chronicle)

above Cal Eastern's flight path whose vertical stabilizer struck just forward of N79992's horizontal stabilizer. As the tail sliced through ONAs' fuselage, the control cables were severed followed by separation of the entire empennage. N79992 nosed over and dove out of control towards the ground, impacting in the middle of Doolittle Drive, a road that runs parallel to the Oakland Airport and onto the island of Alameda. The tail section floated down aimlessly and landed in the marshland next to the road, 100 yards north of the impact area.

Traffic was light on Doolittle Drive, but the wreckage dropped in front of Robert Leibel, 20, and his car struck the debris. He was severely burned. In another car, Norman Barnes and his son were able to stop short of hitting the crashed plane and escaped injury. The paint on their auto was scorched. Frank Ormerod, his wife, and daughter were also able to stop their car, which was showered by debris.

Bob Way, an employee at the Oakland Naval Medical Supply Center, was driving 10 members of the Junior Air Force section of Oakland's Junior Hi-Y, who were riding in the back of his pickup truck. Young Don Gaskin Jr. was quoted in the Nov. 18, 1951, *San Francisco Chronicle*, "First a small piece of the tail fell off, then the whole tail broke, but the heavy plane beat it to the ground, catching fire on the way down.

"When it hit, fire and smoke went up like an atomic bomb. Then red hot metal shot all around us like bullets. When it hit, I screamed and dropped down in the truck, but it didn't save me. A piece of the plane tore off one fender of the truck. Then a streak of flame shot out at us and burned the paint off the back of the truck. That was the one that burned me and the other boys." All occupants of the pick-up were treated for burns and shock.

On board the Cal Eastern DC-4, Captain Shope immediately took the controls after the collision. The impact had spun the plane from a northerly heading to the south and tore off the tail (rudder and vertical stabilizer) 20 inches above the center rudder hinge. Shope was able to maintain control of the aircraft above 160 knots and requested permission to land on Oakland's Runway 9R with crash trucks standing by. Since all available airport emergency equipment was then en route to the Overseas National crash, Shope was diverted to San Francisco. Although he landed "hot," at 140 knots, the crew and aircraft survived to fly again.

The Aftermath

The wreckage of Overseas National's N79992 burned for hours and even threatened a gasoline pipeline supplying the Naval Reserve Air Base on the field. Once the fire was out, the twisted mass of metal was pushed off the road into a parking lot. Examination of the ONA tail revealed a 20-foot piece of antenna wire and paint marks from Cal Eastern's DC-4. Two other pieces of Cal Eastern's tail were found in the vicinity of the crash site. Both aircraft were operating with pilots "under the hood" simulating instrument flying conditions. Fabric covered the left windscreen and side windows, thus requiring the pilot to fly completely blind. This also limited the check pilot's view out the left side of the cabin. Under existing aviation regulations, neither flight was required to be in contact with the field's tower while approaching the Oakland radio range station. Overseas National's rules stated that their check pilots should look for traffic out the astrodome while making turns, but allows them to be in the cockpit while flying straight and level. Even if the third captain had been in the astrodome, it is unlikely that he would have seen the Cal Eastern DC-4 approaching from below and to the rear. After the accident, Cal Eastern immediately changed its company policy to include a third crew member to serve as safety observer on instrument check flights.

The CAB filed a report of violations against the captains of each aircraft, faulting them for failing to see the other aircraft, and Shope was reprimanded for not having a third pilot on the flight deck watching for traffic. As any pilot will tell you, it all boils down to, "See and Be Seen."

Coincidentally, three additional Overseas National captains were scheduled to be on board for flight checks but failed to arrive on time and the flight left without them. How lucky can you get?

N4002B's Subsequent Career And The Crash Site

N4002B was repaired and subsequently sold to TACA, registered YS-04C. During the opening days of 1952, RANSA acquired and re-registered it TI-1046C. On April 4, 1952, Transocean Airlines bought the plane and its new registration became N4664S. Nine years later, in October 1963, Lineas Aereas Paraguayas obtained title to the DC-4 and re-registered it ZP-CAR. One year later, its registration changed once again, this time sold to Paranese, a Brazilian company. Reports state that it was stored at Sao-Paulo and eventually broken up by 1980.

Thousands of cars pass directly over N79992's crash site each day. Probably less than 1 in 10,000 know that an aviation accident occurred at that location, or that three men lost their lives there. Little has changed. The road is still in the same place and the site where the tail section landed is still a marsh. The parking lot next to where N79992 came to rest is now gone. This entire waterfront area has been redeveloped into a park complete with boat launch ramp. A soil berm has been erected between the water and the road, but one can still locate the crash site through pictures. The road has been paved over numerous times, but that will never hide the scars where a DC-4 and its crew came to rest in the Winter of 1951.

WC

Above: *After the fire was put out, the fuselage and wing sections were pushed off the road and into a parking lot. The tail section can be seen in the background sitting right side up.*
(Bob Hofford)

Civil Aeronautics Board photo showing the relative proximity of the two aircraft at the time of impact. (CAB)

A police officer examines the clipped tail of Cal Eastern's DC-4 at San Francisco Airport. (San Francisco Chronicle)

83

Flying Tiger Freighter Impacts Coastal Hills Shortly After Takeoff

By Nicholas A. Veronico

Rain, low clouds, fog and mist covered the San Francisco Bay Area in the late evening and early morning hours of Dec. 23-24, 1964. The day would dawn as Christmas Eve, but for three families this day would become a sad anniversary. The official weather called for scattered clouds at 700 feet under an 1,800-foot overcast ceiling. Visibility was six miles in light rain.

Flying Tiger Line Inc. Flight 282 was scheduled to depart Dec. 23, at 9 p.m., on a domestic cargo trip from San Francisco to New York (JFK). The trip would be flown with Lockheed 1049H N6915C, msn 4812. The scheduled departure time came and went as the company did not have a flight engineer available. A request was made to the Los Angeles, California, station and Flight Engineer Paul M. Entz, age 37, deadheaded on the next flight into San Francisco, arriving at 11:15 p.m. He was an experienced crew member having 4,113 hours of flight time, nearly 93 percent in the 1049H. Entz, having given up Christmas with his family, joined Captain Jabez A. Richard, 49, and Co-pilot Daniel W. Hennessy, 33, in preparing for the flight. Richard joined Tigers on Dec. 4, 1950, and had accumulated 14,911 hours, of which 3,942 were in Lockheed's Super Constellation. Hennessy went to work for the carrier on April 24, 1955, and had accumulated 3,636 hours with 1,277 in type. He was a Korean War veteran and his wife Lorie was a former Flying Tiger stewardess.

The crew had filed an Instrument Flight Rules (IFR) flight plan from San Francisco to Sacramento, then to JFK following Victor Airways, all at 11,000 feet. Flight time was expect-

---

SA-382
CIVIL AERONAUTICS BOARD
File No. 1-0064

AIRCRAFT ACCIDENT REPORT

ADOPTED: June 2, 1966
RELEASED: June 8, 1966

THE FLYING TIGER LINE INC.
L-1049H, N6915C,
SAN FRANCISCO INTERNATIONAL AIRPORT
SAN FRANCISCO, CALIFORNIA
DECEMBER 24, 1964

SYNOPSIS

A Flying Tiger Line Inc., L-1049H, N6915C, operating as Flight 282, crashed on Sweeney's Ridge, approximately 4.3 miles west-southwest of the San Francisco International Airport at approximately 0031:30 P.s.t. The three crewmembers were fatally injured. There were no passengers. The flight, scheduled as domestic cargo from San Francisco to John F. Kennedy International Airport, Jamaica, New York, had just departed Runway 28L at 0028. At 0031:20 the departure controller advised that they were left of course, and within seconds, ". . . the target stopped, bloomed, and disappeared from the radar scope." All attempts to contact the flight following the target disappearance were unsuccessful.

The Board determines that the probable cause of this accident was that the pilot, for undetermined reasons, deviated from departure course into an area of rising terrain where downdraft activity and turbulence affected the climb capability of the aircraft sufficiently to prevent terrain clearance.

1. INVESTIGATION

1.1 History of the Flight

The Flying Tiger Line Inc., L-1049H, N6915C, Flight 282, was a scheduled domestic cargo flight from the San Francisco International Airport (SFO), San Francisco, California, to the John F. Kennedy International Airport (JFK), Jamaica, New York. It was originally scheduled to depart at 2100 1/ 23 December, but the flight was delayed because of the non-availability of a flight engineer. An engineer obtained from Los Angeles, arrived in San Francisco at 2315, and the flight departed at 0028, 24 December.

The original flight plan was IFR: requested altitude - 11,000 feet; route of flight was San Francisco direct to Sacramento, Victor Airways to JFK; estimated time en route - 9 hours 16 minutes; fuel aboard - 11 hours 3 minutes; alternate airport - Newark, New Jersey. Gross weight of the aircraft was 142,073 pounds 2/ of which 29,000 pounds (5,000 gallons) was fuel and 41,078 pounds was cargo.

1/ All times herein are Pacific standard, based on the 24-hour clock.
2/ Maximum gross weight for takeoff from runway 28L at SFO for an L-1049H was 142,100 pounds.

84

ed to be 9 hours 16 minutes, but the plane carried enough fuel (5,000 gallons or 29,000 pounds) for 11 hours 3 minutes aloft. Its alternate was Newark, New Jersey. Cargo for the flight weighed 41,078 pounds — bringing the total weight to 142,073 pounds — only 27 less than its maximum allowable gross takeoff weight.

San Francisco Ground Control granted N6915C permission to taxi towards Runway 28L at 12:13 a.m. Winds were 15 knots with gusts to 23 from 210 degrees, creating a strong crosswind on takeoff. The Clearance Delivery Controller granted the flight its IFR flight plan indicating that the departure would be along the 287-degree radial also known as the "Golden Gate One" departure. From Runway 28, the Golden Gate One departure takes an airplane through a "gap" between the San Bruno Mountains on the bay side of the peninsula and the coastal foothills to the west. Flight 282, callsign "Tiger Niner Fifiteen," read back the clearance including the climb-out on the 287-degree radial, at 12:21 a.m.

Tiger Niner Fifteen advised the tower that they were ready for takeoff six minutes and 45 seconds later. At 12:28, Flight 282 reported to the tower that they were "rolling." The tower controller noted that the aircraft passed in front of the tower and became airborne at 12:30 a.m.

Witnesses on the ground later said that Tiger Niner Fifteen veered to the right and then made a steeper turn to the left before leveling its wings and continuing to climb. They said the plane entered the clouds wings level. These turns were probably course corrections to return to the correct heading after encountering strong crosswinds.

The tower controller instructed the flight to contact Departure Control at 12:30:22 a.m. Thirty seconds later, Flight 282 made contact and said, "Departure, Tiger Nine One Five, you got us— ah, over?" Departure Control acknowledged the transmission five one-hundredths of a second later and advised the flight to report its progress to 11,000 feet in 1,000-foot increments. The crew responded at 12:31:05, "Roger, how do you have us tracking toward the, ah, gap?" The controller switched radar modes, from 30- to 10-mile scan setting, and asked the crew to report its altitude. They replied, "Nine Hundred" at 12:30:16. Departure Control quickly responded by saying, "...you're left of course of the San Francisco two-eight-seven (radial)."

Flight 282 did not respond, nor did it answer any subsequent calls. The Departure Controller said, "...within seconds after the second transmission, the target stopped, bloomed, and disappeared from the radar scope."

The flight lasted a mere 3 minutes and 30 seconds; it was estimated that the plane crashed at 12:31:30 a.m., Dec. 24, 1964. The Constellation had indeed veered left of its intended course and struck Sweeney's Ridge 4.3 miles from the San Francisco International Airport bearing 257 degrees (roughly 2.5 miles left of its intended course). The left wing struck the east side of Sweeney's Ridge at 840 feet above sea level, cartwheeling the fuselage into the hill at 860 feet, narrowly missing a U.S. Coast Guard radio station. The plane exploded and momentum carried parts up and over the 925-foot summit. The CAB report stated that all four engines were producing power and that none of the control surfaces had separated in flight. Both the ADF and VOR receivers were set correctly.

The Connie door rests on Sweeney's Ridge. The door and other large pieces can still be found in the thick underbrush.
(Ian E. Abbott)

The Constellation was carrying 41,000 pounds of "general merchandise" and electronics equipment plus 136 pounds of Christmas mail. It was reported that hundreds of "brightly colored women's scarves were found scattered about the area."

Determining The Cause

Local Sheriff Earl B. Whitmore was quoted in Dec. 24, 1964's *San Mateo Times,* "Had it been 20 feet higher, it might have cleared the ridge." A little bit of an exaggeration but, as always, had it been 100 feet higher it would indeed have cleared the ridge.

A number of factors contributed to this crash. First, the heavily loaded Constellation faced winds and crosswinds from 30 to 35 knots with gusts to 45 knots. As the plane few closer to Sweeney's Ridge, it would encounter turbulence and possibly severe downdrafts.

Second, false VOR indications may have been presented to the crew by a faulty radio transfer switch assembly (or

Right: *Another view of the Connie door showing the outer fuselage skin and locking handle.*

Below: *Authors Veronico, left, and Davies display a number of artifacts found at the Connie crash site. Veronico holds the fuel tank cap and the blade antenna can be seen leaning against the door.*

(Both Ian E. Abbott)

more precisely, the deviation indicator transfer switch) that displays readings from co-pilot's VOR onto the pilot's. The CAB flew a similar "ghost ship" over the crash route in an effort to determine a possible cause, but it was a report from another Flying Tiger Connie that experienced a VOR deviation. That led investigators to look at the crash as a possible contributing factor.

In addition, the co-pilot's request for tracking information may have been an indication of a navigational aid problem or failure. In an effort to ensure this type of accident did not happen again, Flying Tiger Line inspected all radio transfer switch assemblies in its Connie fleet. Three aircraft were found to have faulty switches. Subsequently, the FAA issued an Airworthiness Directive (AD) requiring that each switch be pulled and inspected before it had accumulated 300 hours in service.

Third, the aircraft executed a 55 degree left turn after takeoff. This will never be explained — considering the departure Tiger Niner Fifteen was approved for calls for a slight veering to the right. The CAB report stated, "Since no reason is apparent why the left turn would not have been displayed on the instrument panel, the Board concludes that the crew apparently failed to refer to the total instrument portrayal in the cockpit."

The probable cause for the crash was cited as, "...the pilot, for undetermined reasons, deviated from departure course into an area of rising terrain where downdraft activity and turbulence affected the climb capability of the aircraft sufficiently to prevent terrain clearance."

The Crash Site Today

Sweeny's Ridge is sometimes known as the "Plymouth Rock of the West Coast." From here, explorers of the Spanish Portola expedition had their first views of San Francisco Bay in November 1769. The area of the crash was a U.S. Coast Guard reserve off limits to the public. When the Coast Guard decommissioned the site it was dedicated as a part of the Golden Gate National Recreation Area on May 8, 1984. This location offers tremendous unobstructed views of both the ocean and San Francisco Bay, and is home to mule deer, coyotes, bobcats and mountain lions.

Problems quickly arose when attempting to locate this crash site. The official coordinates were off — way off. The CAB's position put the wreck below 600 feet in elevation and the plane could not have crashed there. A study of aerial photographs was made available by Tom Northop and Pacific Aerial Surveys of Oakland, California. Northrop was able to locate aerial photos taken two years after the crash that still showed the burnt foliage around the wreck site. This started the expedition off in the right direction.

After viewing the photos, we decided to put the new "TOPO!" CD-ROM of USGS maps to the test. The product bills itself as "The Ulitmate Tool For Planning Outdoor Activities Around The Bay Area." After reading the manual and quickly installing the software, the crash coordinates from the CAB report were entered. Using the software, coupled with the aerial photographs, we were able to plot the crash site and determine the correct topo, San Francisco South 7.5 Minute Quadrangle Map. Viewing the topo on the screen, the Coast Guard reservation and its buildings were clearly marked. The map was printed on a laser printer and the group headed for the site.

We proceeded to Skyline Junior College, parked in the southern-most parking lot near the trail head and the school's maintenance yard. Although Sweeney's Ridge is not marked on the topo, it is part of the Golden Gate National Recreation Area and is listed in their publications. The group started up the well marked and maintained trail. At the top of the trail, approximately one-half mile, the trail forks. Heading to the right to the top of Sweeney's Ridge, we walked another one-tenth of a mile before turning off the trail to the east. From this point, the San Francisco Jail could be seen below the hillside. All of the Coast Guard buildings were torn down in the 1950s.

The first pieces of wreckage were found 50 yards down the face in an area of thick ground cover. Be aware that parts of the old Coast Guard reservation and radio station have been mixed in with the Connie's remains. This material is primarily electrical components, street lights, and wood. Many people have ventured into this area not knowing its significance. Dozens of beer cans and bottles litter the area.

Once clear of the recent debris, a number of large pieces can be quickly located. One or two flaps, a door complete with latch- handle- and locking-mechanism, and the radio blade antenna were found. A portable am/fm radio that was brutally twisted bore the scars of the crash including a small pebble imbedded in its face. One of the wing fuel cell caps, still wearing a top coat of red paint, turned up. Dozens of pieces with Lockheed part numbers were found, making the wreck even more interesting. Portions of Flight 282's cargo still can be found with a little digging. Blue beads from the costume jewelry plus gloves and scarves still wrapped in plastic — some with melted or scorched packaging — were found in large numbers.

Flying Tiger N6915C is one of the most accessible crash sites with a walking trail less than 25 yards from the point of impact. Its proximity to the coast, coupled with the site's thick vegetation, calls for layers of long-sleeve shirts and good eye protection. No poison oak was noticed during our visit. Sweeney's Ridge is a popular hiking area, yet hundreds of pieces still litter the site more than 30 years later.

This wreck was also the acid test of "TOPO!" Did it work? Outstanding is the answer. After returning to the office, the precise location was entered into the data base and the correct coordinates were provided: Latitude: 37°37'28" by Longitude 122°27'45".

WC

This gear section was found partially buried on the hillside. Rumors persist that one of the Connie's engines remains on the far side of Sweeney's Ridge.
(Ian E. Abbott)

This section, possibly from the flaps or wing area, is nearly three feet long. It was found 100 yards from the top of the hill.
(Ian E. Abbott)

CRASH SHORTS: A TRIO OF C-46 CRASHES

9,046-foot Panther Peak Grabs California Airways C-46 N1240N

Curtiss C-46F-1-CU, msn 22404, was accepted by the U.S. Army Air Force in July 1945, and assigned serial number 44-78581. This aircraft was one of six C-46s leased from the U.S. Air Force by Alaska Airlines on June 8, 1948. The plane was registered N1240N.

In February 1950, N1240N was returned to the U.S. Air Force. It subsequently was leased to Slick Airways on Feb. 21, 1950, along with 14 other C-46s. Slick was a major post-war cargo airline that began life as a contract cargo carrier in January 1946. These flights primarily were freight runs for the U.S. military. Slick's primary routes were between California, Texas, and the Eastern seaboard. The carrier eventually would operate a fleet of 35 Curtiss Commandos. On Nov. 27, 1950, engine overhaul paperwork was filed with the FAA showing the aircraft's total time of 2,157 hours.

Slick returned title of the aircraft to the U.S. Air Force on Sept. 28, 1954, but continued to operate the plane. On June 1, 1955, Slick purchased title from the Air Force for $70,000.

On Sept. 9, 1957, the plane was sold to Axxico Airlines Inc. of Miami, Fla. Although owned by Axxico, it was operated by and flown in the colors of California Airfreight. An airworthiness certificate dated Feb. 14, 1958, showed the aircraft's total time of 15,644 hours.

On Jan. 13, 1959, the plane struck the south side of 9,046-foot Panther Peak in Sequoia National Park. Two died in the crash, just 200 feet below the summit of Panther Peak. **WC**

Top: *Non-scheduled Slick Airways flew cargo between California, Texas, and the eastern seaboard. Here, N1240N wears Slick's colors somewhere in the Mid-western United States in the early 1950s.*
(William T. Larkins Collection)

Above: *N1240N's last paint scheme in the colors of California Airfreight.* (William T. Larkins Collection)
Below: *Curtiss C-46 N1240N came to rest just 200 feet below the summit of 9,046-foot Panther Peak in the Sequoia National Park in Central California. Numerous large pieces remain at the site today. Jerry Boal lends scale to the C-46A's tail.* (G. Pat Macha Collection)

Another 200 Feet And...
The Wreck Of Aerovias Contreras XA-GOT "Ensenada"

Another 200 feet and she would have made it. Unfortunately, Aerovias Contreras Curtiss C-46A crashed into the mountains east of San Diego, Calif., near the Mexican border on Aug. 18, 1953.

The plane originally was built as a C-46A-1-CU "Commando" for the U.S. Army Air Corps. Assigned serial number 41-5201, msn 26403, it was accepted on Dec. 29, 1942, and delivered to the North Atlantic Wing of the Air Transport Command on Feb. 22, 1943. After the end of the war, 41-5201 was sent to Ontario, Calif., for storage and subsequently sold by the Reconstruction Finance Corp. The Commando was purchased by Aerovias Contreras, S.A., a Mexican freight hauler headed by Manual Contreras. The company operated one other C-46, msn 26796, which was registered both XA-HIO and XA-JEB. Aerovias Contreras christened XA-GOT "Ensenada."

The plane flew south of the border for almost six years before striking the hills east of San Diego. **WC**

Above: *Curtiss C-46A-1-CU was flown on non-scheduled cargo runs between border towns and the interior of Mexico. Here, XA-GOT basks in the sun at San Diego.* (William T. Larkins Collection)

Opposite Page, clockwise from top right: *Jumbled wreckage marks the final resting place of "Ensenada"; XA-GOT struck the mountain just below the peak in rugged, steep terrain; Aircraft wreckage with registration visible on the wing, (note the steep terrain); The massive box spar of the Curtiss wing in the hills east of San Diego, Calif.* (All G. Pat Macha Collection)

Coordinates for the crash site of XA-GOT are:

32°35'N 116°49'W

Fist Fight Before The Crash Was Not A Factor
The Crash Of Standard Airlines Curtiss C-46E N79978

Curtiss C-46E N79978 was the eighth of 17 Curtiss C-46E-1-Cs built at the company's St. Louis, Mo., factory. The "E" model Commando's most distinguishing external feature was its stepped windscreen — much like other conventional transports. It also model-featured Pratt & Whitney R-2800-75 engines of 2,000 horsepower each and a single, portside cargo door.

This aircraft was delivered to U.S. Army Air Corps on July 17, 1945 at Sedalia, Mo., as serial number 43-47410, msn 2936. The plane stayed within the continental United States until assigned to the War Assets Administration at Clinton, Oklahoma, on Feb. 9, 1947. Sold to Trans National Airlines on April 5, 1948, and registered N79978, the plane was operated in Standard Airlines colors and flown on non-scheduled passenger flights. N79978 crashed on July 13, 1949, and had flown a total of 1,515 hours.

The plane went down at 7:52 a.m. at 1,940 feet on the south slope of Chatsworth Peak. The peak is located in the Santa Susana Pass, 30 miles west of Los Angeles and one mile north of the Chatsworth Reservoir. There were 36 fatalities, yet 13 others survived.

Detective Sgt. A.M. McDaniel of the Los Angeles County Sheriff's Aero Detail said of the accident in his July 25, 1949, report, "Investigation revealed this plane to have been flying a southwesterly course in level flight with the gear down. The right wing tip brushed the side of the hill, pulling the plane around 90 degrees. It hit the ground and bounced through the air approximately 300 feet to its final position, headed at a right angle to its original path of flight. In the wreckage was found an altimeter registering 1,940 feet. This altitude was later checked with a calibrated altimeter and found to be correct. This crash location is on course with the normal instrument letdown procedure for Lockheed Airport, but the altitude normally used at this point is 4,000 feet."

Newspaper headlines the following morning of July 14, 1949, exclaimed, "Investigators Doubt Fight in Plane Caused Crash That Cost 35 Lives." Stewardess Charolette Grenander, who survived the crash with a broken leg, cuts and bruises, said the fight between two passengers occurred about 45 minutes before the crash and was settled within 15 minutes. The man who started the fight was identified as Frank Conway of Albany, N.Y., who died in the crash. The stewardess was unable to identify the second party. She related that the plane's pilot, Captain Roy G. White, had come back to ensure that the fight had been settled.

The CAB determined that the fight was of no significance in the accident. The plane was too low on approach and at that geographic location, and should have been at 5,000 feet.

WC

Above: *N79978 on the side of Chatsworth Peak in the Santa Susana Pass. The crash site is located 30 miles northwest of Los Angeles and one mile north of the Chatsworth Reservoir. There were 36 fatalities, and 13 survivors.* (G. Pat Macha Collection)
Opposite Page: *Beautiful night study of C-46E N79978 by William T. Larkins. Note the stepped windscreen, featured only on 17 "E" models built in St. Louis, Missouri.*

APPENDIX I

Jet And Turbo-Prop Airliner Crashes

This list is intended as a starting point for research purposes only. It does not list every aircraft of a particular type that has crashed. The authors/publishers make no guarantees that wreckage still exists at these sites. Wreck Chasers are cautioned to do their research and be prepared before setting out on an adventure.

Type	MSN	Reg.	Date	Airline	Nearest Major City
Hawker Siddeley HS 748					
HS 748	1536	G-ARMV	07/11/65	Skyways Coach Air	Lympne, Kent, UK
HS 748	1539	LV-HGW	02/04/70	Aerolineas Argentinas	Corrientes, Argentian
HS 748	1540	LV-HHB	04/15/76	Aerolineas Argentinas	Neuquen, Argentina
HS 748	1546	LV-HHH	01/20/71	Aerolineas Atgentinas	Sarmiento, Argentina
HS 748	1560	G-ASPL	06/26/81	Dan-Air	Nailstone, England
HS 748	1568	HS-THB	04/27/80	Thai Airways	Bangkok, Thailand
HS 748	1580	YC-C-AMY	08/20/68	Linea Aeropostal Venezolana	Maturin, Venezuela
HS 748	1582	C-GDOV	01/12/89	Bradley Air Services	Dayton, Ohio
HS 748	1586	RP-C1029	05/10/75	Philippine Airlines	Manila, Philippines
HS 748	1590	RP-C1028	02/03/75	Philippine Airlines	Manila, Philippines
HS 748	1598	XA-SEV	01/06/72	SAESA	Chetumal, Mexico
HS 748	1617	C-GQTH	11/10/93	Northland Air Manitoba	Sandy Lake, Ontario, Canada
HS 748	1618	C-GQSV	12/03/88	Air Creebec	Waskaganish, Canada
HS 748	1620	PK-IHE	01/09/93	Bouraq Indonesia Airlines	Surabaya-Juanda, Indonesia
HS 748	1625	PP-VDN	06/17/75	VARIG	Pedro Afonso, Gioias, Brazil
HS 748	1632	PP-VDU	02/09/72	VARIG	Porto Alegre, Brazil
HS 748	1633	PK-IHK	02/09/77	Bouraq Indonesia Airlines	Ujung Pandang, Indonesia
HS 748	1635	C-GEPH	XX/29/81	Eastern Provincial Airways	Sydney, Nova Scotia
HS 748	1637	RP-C1015	06/26/87	Philippine Airlines	Crashed Mt. Pugo, Baguio, Philippines
HS 748	1643	RP-C1022	04/21/70	Philippine Airlines	Manila, Philippines
HS 748	1673	XA-SAB	07/27/73	SAESA	Acapulco, Mexico
HS 748	1683	HC-AUE	01/20/76	TAME Ecuador	Andes Mts. near Guayaquil, Ecuador
HS 748	1704	FAC-1103	01/09/74	SATENA	Florencia, Columbia
HS 748	1705	FAC-1104	08/07/83	SATENA	Pasto, Columbia
HS 748	1789	C-GFFA	09/15/88	Bradley Air Services	Ottawa, Canada
BAe Jetstream 31					
Jetstream 31	623	N823JS	02/09/88	Jetstream Int'l Airways	Springfield, Ohio
Jetstream 31	683	N165PC	03/12/92	CCAir	Knoxville, Tennessee
Jetstream 31	706	N334PX	12/01/93	Express Airlines	Hibbing, Minnesota
Jetstream 31	742	N331CY	05/26/87	Air New Orleans	New Orleans, Louisiana
Jetstream 31	776	N410UE	12/26/89	North Pacific Airlines	Pasco, Washington
Jetstream 31	787	N131CA	04/20/93	Westair Commuter Airlines	Merced, California
Bristol Britannia 102					
102	12903	G-ANBB	09/01/66	Britannia Airways	Ljublijana Airport, Yugoslavia
253 C Mk1	13448	G-BRAC	02/16/80	Red Coat Air Cargo	Boston, Massachusetts
Canadair CL-44-6/CC-106					
	1	HK-1972	02/22/75	Aerocondor Columbia	Alto de Toledo, Columbia
CL-44-6/CC-106	12	OB-R1104	08/27/76	Aeronaves del Peru	Lima, Peru

Type	MSN	Reg.	Date	Airline	Notes
CL-44D4-2	22	N453T	03/21/66	Flying Tiger Lines	Norfolk, Virginia

CASA 235

Type	MSN	Reg.	Date	Airline	Notes
CASA 235-10	14	PK-MNN	10/18/92	Merpati Nusantara Airlines	Mt. Papandayan, Western Java

Convair 580

Type	MSN	Reg.	Date	Airline	Notes
Convair 580	23	N73130	03/05/67		Marseille, Ohio
Convair 580	56	LN-PAA	09/08/89		Hirtshals, Denmark
Convair 580	83	N90858	06/29/72	North Central Airlines	mid-air with Air Wisconsin DHC-6 N4043B
Convair 580	98	N25278	09/19/91	Canair Cargo	Burlington, Vermont
Convair 580	157	YV-84C	05/28/85	AVENSA	Cabimas, Venezuela
Convair 580	180	ZK-FTB	07/31/89	Fieldair	Auckland, New Zealand
Convair 580	380	N4825C	07/24/78	North Central Airlines	Kalamazoo, Michigan
Convair 580	384	N5832	06/07/71		New Haven, Connecticut
Convair 580	386	N5825	01/06/69	Allegheny Airlines	Bradford, Pennsylvania
Convair 580	410	N5802		Allegheny Airlines	Bradford, Pennsylvania
Convair 580	452	N67701	12/12/63	Union Oil Co. of California	Midland, Texas

Convair 640

Type	MSN	Reg.	Date	Airline	Notes
Convair 640	9	N862FW	02/09/92		Cape Skirrinng, Senegal

DHC-6 Twin Otter

Type	MSN	Reg.	Date	Airline	Notes
DHC-6-100	5	N7705	05/25/73	Air Central	Savoy, Illinois
DHC-6-100	6	VH-TGR	04/28/70	Trans Australia Airlines	Kainatu, New Guinea
DHC-6-100	13	N4043B	06/29/72	Air Wisconsin	mid-air with North Central Convair 580 N90858, Neenah, Wisconsin
DHC-6-100	18	XA-BOP	09/09/78	Lineas Aereas del Centro	Mexico City, Mexico
DHC-6-100	19	C-FAJB	12/16/72	Airwest Airlines	Vancouver Island, British Columbia
DHC-6-100	30	I-CLAI	03/11/67	Aeralpi	Mt. Visentin, Borga Prigola, Italy
DHC-6-100	31	D-IDHC	05/27/67	General Air	Heligoland, Germany
DHC-6-100	34	C-FAUS	11/11/84	Labrador Airways	Mealy Mtns., Goose Bay, Newfoundland
DHC-6-100	38	N2711H	09/09/69	Interior Airways	Sagwan, Alaska
DHC-6-100	39	BR-GCP		Guyana Airways	Georgetown, Guyana
DHC-6-100	56	67-056	07/11/72	Royal Norwegian Air Force	Harstad, Norway
DHC-6-100	58	CF-WGE-X	06/13/87	Athabaska Airways	Maudsley Lake, Canada
DHC-6-100	63	67-063	10/29/90	Royal Norwegian Air Force	Honningsvaag, Norway
DHC-6-100	67	CF-AWF	09/22/76		Mosher Creek, Bella Coola, British Columbia
DHC-6-100	68	N202RH	06/17/88	Samoa Aviation	Tau Manua, American Samoa
DHC-6-100	70	PK-NUC	02/28/73	Merpati Nusantara Airlines	Nabire, Indonesia
DHC-6-100	76	PK-MAM	05/25/87	Missionary Aviation Fellowship	Ilaga, Indonesia
DHC-6-100	83	N6383	01/09/75	Golden West Airlines	mid-air Whittier, Calif., Cessna 120 N11421
DHC-6-100	84	P2-RDE	02/28/78	Talair	Garaina, Papua New Guinea
DHC-6-100	85	ST-ADB	03/18/75	Sudan Airways	Khartoum, Sudan
DHC-6-100	95	CF-QBT	03/06/93	Kenn Borek Air	Baffin Island, NWT, Canada
DHC-6-100	100	C-GPBO	12/01/77	Airwest Airlines	Satuma Island, British Columbia
DHC-6-100	102	9N-RF9	02/27/70	Nepalese Royal Flight	Jomson, Nepal
DHC-6-100 (-300)	105	N127PM	02/21/82	Pilgrim Airlines	Providence, Rhode Island
DHC-6-200	122	C-FWAF	09/30/79	West Coast Air Services	Porpoise Bay, British Columbia
DHC-6-200	125	N4048B	10/13/78	Artic Guide	Barrow, Alaska
DHC-6-200	131	N187SA	04/18/90	Aeroplerlas	Contadora, Panama
DHC-6200	141	N141VP	04/22/92	Perris Valley Aviation Services	Perris, California
DHC-6-200	142	CF-MHU	09/30/75	Northern Thunderbird Air	Klua Tan Tan, British Columbia
DHC-6-200	159	N996SA	01/19/88	Mountain Air Cargo	Charlotte, North Carolina
DHC-6-200	166	OH-KOA	02/05/73	Kar-Air Finland	Pudasjarui, Finland
DHC-6-200	175	N558MA	07/15/60	New York Airways	New York, New York
DHC-6-200	183	N103AQ	05/16/82	Kodiak Aviation	Hooper Bay, Alaska
DHC-6-200	192	N3257	10/11/85	Mountain Air Cargo	Homer City, Pennsylvania
DHC-6-200	195	N7267	07/20/88	Fairways Corp.	Dulles, Washington, D.C.
DHC-6-200	196	N563MA	09/06/77	Alaska Aeronautical Industries	Mt. Iliamna, Alaska
DHC-6-200	203	N958SM	11/09/70	Mississippi Valley Airlines	La Crosse, Wisconsin
DHC-6-200	213	C-FGIK	12/17/83	Inuvik Coastal Airways	Paulatuk, NWT, Canada
DHC-6-200	215	C-FAIV	09/03/78	Stanley Park	Vancouver, British Columbia
DHC-6-200	227	VH-PAQ	02/13/79	Pan-Air	Yarra Creek, King Island, Tasmania
DHC-6-200	230	T-87	08/07/77	Lineas Aereas del Estado	Bariloche, Comodoro Rivdavia, Argentina

Type	MSN	Reg.	Date	Airline	Nearest major city
DHC-6-300	236	N76214	07/13/78	Coastair	Gander, Newfoundland
DHC-6-300	246	CF-QDG	08/23/78	Bradley Air Services	Frobisher Bay, Canada
DHC-6-300	248	N76GC	06/18/86	Grand Canyon Airlines	near Grand Canyon Airport, Arizona
DHC-6-300	251	N389EX	04/28/74	Air America	Laos
DHC-6-300	252	HK-1710W	12/19/79	Am Son Drilling Co.	Bogota, Columbia
DHC-6-300	256	H4-SIB	09/27/91	Solomon Islands Airlines	Honiara, Solomon Islands
DHC-6-300	258	F-OGFE	12/21/72		Crashed off St. Martin
DHC-6-300	262	N101AC	12/12/76	Atlantic City Airlines	Wildwood, New Jersey
DHC-6-300	278	CF-ABW	12/21/77	Kenn Borek Air	Namisivik, NWT, Canada
DHC-6-300	294	N8061V	12/03/79	Evergreen Helicopters	Nome, Alaska
DHC-6-300	302	9N-ABB	07/05/92	Royal Nepal Airlines	Jumia, Nepal
DHC-6-300	326	N5662	07/25/72	Air America	Laos
DHC-6-300	333	VP-FAP	01/24/77	British Antarctic Survey	Antartica
DHC-6-300	341	8Q-GIA	02/07/87	Inter Atoll Air	Maldives Islands
DHC-6-300	345	PK-NUE	09/15/89	Merpati Nusantara Airlines	Manokwasi-Benuti, Indonesia
DHC-6-300	346	PK-NUG	08/12/85	Merpati Nusantara Airlines	Mulia, Iran Jaya, Indonesia
DHC 6 300	349	C-FWAB	06/06/90	Ptarmigan Airways	Thistle Lake NWT, Canada
DHC-6-300	354	C-FCSV		Canadian Dept. Trans	Galt, Ontario, Canada
DHC-6-300	361	N361V	02/15/83	Sierra Pacific Airlines	Sun Valley, Idaho
DHC-6-300	363	CP-1018	05/19/87	YPF Bolivianos	El Trompillo, Santa Cruz, Bolivia
DHC-6-300	366	P2-RDW	07/21/89	Air Guadeloupe	Porega, Papua New Guinea
DHC-6-300	370	9N-ABG	10/15/73	Royal Nepal Airlines	location unknown
DHC-6-300	376	9N-ABH	12/22/84	Royal Nepal Airlines	Bhojpur, Nepal
DHC-6-300	387	N25RM	12/04/78	Rocky Mountain Airways	Steamboat Springs, Colorado
DHC-6-300	391	PK-NUL	06/17/93	Merpati Nusantara Airlines	Nibire, Irian Jaya, indonesia
DHC-6-300	392	9N-ABI	05/03/86	Royal Nepal Airlines	Sanphe Bagar, Nepal
DHC-6-300	393	6V-ADE	12/10/93	Air Senegal	mid-air with YS-11, Dakar, Senegal
DHC-6-300	394	TZ-ACH	06/21/83	Air Mali	Bamako, Mali
DHC-6-300	395	YA-GAZ	03/10/83	Bakhtar Afgan Airlines	Shashgow Ghazni, Afghanistan
DHC-6-300	400	N707PV	10/28/89	Aloha Island Air	Molokai, Hawaii
DHC-6-300	401	N540N	03/12/85	Seair Alaska Airlines	Barter Island, Alaska
DHC-6-300	408	LN-BNM	10/27/93	Wideroes Flyveselskap	Namsos, Norway
DHC-6-300	416	TJ-CBC	10/31/81	Cameroon Airlines	Bafoussam, Cameroon
DHC-6-300	417	C-GKBM	07/14/88	Kenn Borek Air	John Day, Oregon
DHC-6-300	419	HK-2920	10/11/87	Aerojecutivos Columbia	El Poleo, Columbia
DHC-6-300	428	C-GDHA	05/03/76	deHavilland Canada	Monze, Zambia
DHC-6-300	431	N23BC	11/05/79	South Pacific Island Airways	Tau Island, American Samoa
DHC-6-300	439	N75GC	09/27/89	Grand Canyon Airlines	Grand Canyon, Arizona
DHC-6-300	452	TN-ACX	03/12/92	Lina Congo	Estouali, Congo
DHC-6-300	457	HC-BAX	05/22/81	TAME Ecuador	Zumba, Ecuador
DHC-6-300	459	PK-NUY	12/30/87	Merpati Nusantara Airlines	Samarinda-Berau, Indonesia
DHC-6-300	463	C-GNTB	01/14/77	Northern Thunderbird Air	Terrace, British Columbia
DHC-6-300	465	N982FL	01/18/78	Frontier Airlines	Pueblo Memorial Airport, Colorado
DHC-6-300	469	F-OGHD	11/18/78	Air Guadeloupe	Marie Galante Island, Guadeloupe
DHC-6-300	473	PK-NUT	09/16/93	Merpati Nusantara Airlines	location unknown
DHC-6-300	474	PK-NUW	05/23/87	Merpati Nusantara Airlines	Ruteng, Flores Island, Indonesia
DHC-6-300	486	PK-NUP	03/29/77	Merpati Nusantara Airlines	Bainaha Valley, Sulawesi, Indonesia
DHC-6-300	488	PK-NUQ	04/03/90	Merpati Nusantara Airlines	Lebuhanraio, Indonesia
DHC-6-300	497	HK-1910	01/23/85	Aces Columbia	Quibdo-Medelin, Columbia
DHC-6-300	501	XY-AEB	08/12/82	Burma Airways	Mindat, Burma
DHC-6-300	512	XY-AEE	10/08/83	Burma Airways	Lonkin Airfield, Burma
DHC-6-300	526	YV-30C	04/17/78	Linea Aeropostal Venezolana	Vrima, Venezuela
DHC-6-300	536	LN-BNS	04/12/90	Wideroes Flyveselskap	into sea, Vaeroey, Loften Islands
DHC-6-300	540	XY-AEH	09/08/77	Burma Airways Corp.	Monghsa-Keng Tung, Burma
DHC-6-300	541	XY-AEI	08/26/78	Burma Airways Corp.	Papun, South Burma
DHC-6-300	568	LN-BNK	03/11/82	Wideroes Flyveselskap	North Cape, Norway
DHC-6-300	581	5H-MRD	12/20/84	Air Tanzania	Dares Salaam, Tanzania
DHC-6-300	606	HK-2889X	07/04/91	Helicol Columbia	El Yopal, Casnare, Columbia
DHC-6-300	609	HK-2216	12/18/81	Aces Columbia	Sanantero, Columbia
DHC-6-300	622	7P-LAA	07/13/84	Lesotho Airways	Mokhotiong, Lesotho
DHC-6-300	630	C-GTJA	11/01/79	Austin Airways	Big Trout Lake, Ontario, Canada
DHC-6-300	632	C-GTLA	11/23/83	Austin House	Lansdowne House, Thunder Bay, Canada
DHC-6-300	670	5A-DDD	11/30/88	Aero Club Libya	Hamada Oilfield, Libya
DHC-6-300	676	HK-2486	06/11/89	Aerotaca	Caribabre, Columbia

Type	MSN	Reg.	Date	Airline	Nearest major city
DHC-6-300	699	ET-AIL	08/07/89	RRC Air Service	Fugnido, Ethiopia
DHC-6-300	713	HK-2536	11/29/82	Aces Columbia	Cerro Pan de Azucar, Columbia
DHC-6-300	737	TY-BBL	02/24/83	Air Benin	Koko, near Beteron, Benin
DHC-6-300	771	HK-2759	06/06/93	Aces Columbia	El Yopal, Columbia
DHC-6-300	780	HK-2761	04/27/86	Aces Columbia	Saravena, Columbia
DHC-6-300	785	V2-LCT	08/03/86	LIAT	St. Vincent-Amos Vale
DHC-6-300	819	ET-AIQ	06/22/86	Ethiopian Airlines	Dembidollo, Ethiopia

DeHavilland DHC-7

Type	MSN	Reg.	Date	Airline	Nearest major city
DHC-7 102	028	LN-WFN	05/06/88	Wideroes Flyveselskap	Broennoysund, Norway

DeHavilland DHC-8

Type	MSN	Reg.	Date	Airline	Nearest major city
DHC-8-102	061	N819PH	04/16/88	Horizon Airlines	Seattle-Tacoma, Washington
DHC-8-311	210	D-BEAT	01/06/93	Contactair Flugdienst	Paris, France

Dornier 228

Type	MSN	Reg.	Date	Airline	Nearest major city
228-100	4358	D-IFNS	03/26/82	Dornier	Igenhausen, Augsburg, Germany
228-201	8052/0052	VT-EJF	09/24/89	Vuyadoot	Lake Indopur, Pune, India
228-201	8110/0110	B-12238	02/28/93	Formosa Airlines	into sea off Green Island, Taiwan
228-201	8151/0151	B-12298	06/14/93	Formosa Airlines	into sea off Green Island, Taiwan
228-212	8196/0196	F-OHAB	04/19/91	Air Tahiti	into sea off Nuku Hive Island

Embraer Emb.110 Banderiante

Type	MSN	Reg.	Date	Airline	Nearest major city
110C	110004	PT-TBA	10/28/76	Nordeste Linhas Aereas	Petrolia-Pemanbuco, Brazil
110C	110005	PT-TBB	02/03/92	Nordeste Linhas Aereas	Caetite, Bahia, Brazil
110C	110010	PP-SBB	02/08/79	TAM Linhas Aerea Regional	Bauru, Adudos, Brazil
110C	110011	PT-TBD	01/22/76	Transbrazil	Chapeco, Santa Catarina, Brazil
110C	110013	PP-SBC	06/28/84	TAM Linhas Aerea Regional	Sao Pedro d'Aldeia, Brazil
110C	110021	PP-SBE	02/27/75	VASP	Sao Paulo, Brazil
110C	110037	PP-SBJ	03/23/93	TAM Linhas Aerea Regional	Tangara de Serra, Cuiba, Brazil
110P	110063	PT-GJN	06/23/85	TABA	Cuiaba, Brazil
110EJ	110088	PT-GJZ	08/14/84	VOTEC	mid-air with PT-GKL (msn 110107) near Imperatriz, Maranhao State, Brazil
110C	110090	PT-GKA	10/09/85	Nordeste Linhas Aereas	Vitoria, Brazil
110P	110092	PT-GKC	05/22/82	Rio Sul del	Florianopolis, Brazil
110P	110107	PT-GKL	08/14/84	VOTEC	mid-air with PT-GJZ
110P	110130	PT-GKT	01/31/78	Rio Sul del	Eirunepe, Brazil
110P1	110206	HK-2651	09/02/81	El Venado	Paipa, Columbia
110P1	110226	PT-GMS	11/14/88	Wasa Wings	Ilmajoki, Finland
110P1	110275	N731A	10/08/91	BAC Charter	Narssarssuaq, Greenland
110P1	110294	G-HGGS	11/19/84	Euroair Transport	Inverness, Scotland
110P1	110302	HK-2593	08/04/85	Aires Columbia	Mocoa, Columbia
110P1	110314	PT-SCU	11/11/91	Uniao Taxi Aereo	Recife, Brazil
110P1	110330	N95PB	05/06/89	Southern Express	Mt. Pleasant, Tennessee
110P1	110341	HK-2638	01/23/85	Aires Columbia	Buga, Columbia
110P1	110365	N96PB	12/06/84	Provincetown-Boston Airline	Jacksonville, Florida
110P1	110366	HK-2743	--/--/85	Tavina	location unknown, Columbia
110P1	110368	PT-FAW	09/20/90	Sudene	Fera de Noronha, Brazil
110P1	110370	N1356P	03/13/86	Simmons Airlines	near Alpena-Phelps Collins Airport, Mississippi
110P1	110389	N65DA	05/24/88	Atlantic Southeast Airlines	Lawton, Kansas
110P1	110391	G-ZAPE	01/13/93	Titan Airways	Sellafield, Cumbria, England
110P1	110402	ZS-LGP	03/01/88	Bopair	Johannesburg, South Africa

Embraer Emb.120 Brasilia

Type	MSN	Reg.	Date	Airline	Nearest major city
120RT	120019	N219AS	09/19/86	Atlantic Southeast Airlines	San Jose dos Campos, Brazil, on delivery
flight 120RT	120033	F-GEGH	12/21/87	Air Littoral	Bordeaux, France
120RT	120077	N33701	9/11/90	Continental Express	Eagle Pass, Texas
120RT	120218	N270AS	04/05/91	Atlantic Southeast Airlines	Brunswick, Georgia

Fairchild/Swearingen Merlin/Metro

Type	MSN	Reg.	Date	Airline	Nearest major city
SA226AT	AT-005	N960M	04/14/75		Southern Pines, North Carolina
SA226AT	AT-008	N577KA	05/07/86	Career Aviation Academy	Billings, Montana
SA226TC	TC-228	N650S	06/12/80	Air Wisconsin	Valley, Nebraska

Type	MSN	Reg.	Date	Airline	Nearest major city
SA226TC	TC-231	N200KC	08/17/93	Aviation Services	Hartford, Connecticut
SA226TC	TC-240	N63Z	01/30/84	Britt Airways	Terre Haute, Indiana
SA226TC	TC-275	VH-SWO	05/13/80	Skywest Airlines	Esperence, Western Australia
SA226TC	TC-327	N163SW	01/15/87	Skywest Airlines	mid-air with Mooney M.20C N6485U, Salt Lake City, Utah
SA226TC	TC-334E	F-GCPG	11/18/88	Air Littoral	Lepaud-Montlucon, France
SA227AC	AC-449	N30093	12/07/82	Pioneer Airlines	Pueblo, Colorado
SA227AC	AC-457	N68TC	01/19/88	Trans-Colorado Airlines	Durango, Colorado
SA227AC	AC-481	C-GSLB	09/26/89	Skylink Airlines	Terrance, Canada
SA227AC	AC-500	D-IABB	02/08/88	Numberger Flugdienst	Dusseldorf, Germany
SA227AC	AC-545	N342AE	08/25/92	Lone Star Airlines	Hot Springs, Arkansas
SA227AC	AC-622	N622AV	02/19/88	AVAir	Raleigh-Durham, North Carolina
SA227AC	AC-653	N2689E	09/29/87	Horizon Airlines	Twin Falls, Idaho
SA227AC	AC-683	N683AV	02/01/91	Skywest Airlines	Ground collision with USAir Boeing 737-3B7, msn 23310, N388US, at LAX
SA227AC	AC-716	N2721M	01/15/90	Skywest Airlines	Elko, Nevada

Fairchild F-27

Type	MSN	Reg.	Date	Airline	Nearest major city
F-27A	1	HC-ADV	11/07/60	AREA Ecuador	Quito, Ecuador
F-27	6	N2703	01/17/63	West Coast Airlines	Great Salt Lake, Salt Lake City, Utah
F-27	19	PI-C871	03/08/68	Air Manilla Int'l	into sea off Ibajay Village, Manilla-Cebu
F-27B	21	N4904	08/30/75	Wein Air Alaska	Gambell Airport, St. Lawrence Island
F-27A	29	N745L	11/15/64	Bonanza Air Lines	Las Vegas, Nevada
F-27	32	N2707	08/24/63	West Coast Airlines	Calgary, Canada
F-27A	36	N2770R	05/07/64	Pacific Air Lines	San Ramon, California, crew shot in flight
F-27	47	CF-QBL	03/29/79	Quebecair	Quebec City, Canada
F-27B	49	N4905	12/02/68	Wein Consolidated Airlines	Spotsy Lake, Alaska
F-27A	63	CC-CJE	12/09/82	Aeronor Chile	La Serena, Chile
F-27J	113	CF-GND	06/12/68	Great Northern Airways	Resolute Bay, Canada
F-27J	118	CP-1117	06/02/80	Lloyd Aereo Boliviano	Tapecua Hills, Yacuiba, Bolivia
F-27F	123	N27W	05/04/68	Eastex Inc	Rancho-Paisano Airport, Bruni, Texas
F-27F	125	CF-PAP	02/08/69	Stanair Ltd.	Mikaa Lake, Canada
F-27J	127	CP-862	03/16/84	Lloyd Aereo Boliviano	El Pilon Hill, San Borja, Bolivia

Fairchild FH-227

Type	MSN	Reg.	Date	Airline	Nearest major city
FH-227C	517	N380NE	10/25/68	Northeast Airlines	Moose Mountain, Lebanon Airport, N.H.
FH-227B	522	PT-LCS	01/25/93	TABA	Oeiras-Pa, Belem, Brazil
FH-227B	531	N7811M	11/19/69	Mohawk Airlines	Mount Pilot, Glenn Falls, New York
FH-227B	532	F-GGDM	04/10/89	Uni-Air	Valence, France
FH-227B	536	PT-LBV	06/12/82	TABA	Tabatinga, Brazil
FH-227B	541	N7818M	03/03/72	Mohawk Airlines	Albany, New York
FH-227B	546	F-GCPS	03/04/88	Transport AerienTransregional	Fontainebleau, France
FH-227B	556	PP-BUF	03/14/70	Paraense Transportes Aereos	Baja de Guajara Bay, Belem, Brazil
FH-227B	557	N712U	08/10/68	Piedmont Ailines	Charleston, West Virginia
FH-227B	570	PT-ICA	06/06/90	TABA	Altamira, Brazil

Fokker F.27

Type	MSN	Reg.	Date	Airline	Nearest major city
F.27-100	10108	N148PM	01/13/84	Pilgrim Airlines	New York, New York
F.27-100	10112	VH-TFB	06/10/60	Trans Australia Airlines	Mackay, Queensland, Australia
F.27-100	10118	YN-BZF	04/20/85	Aeronica	Kulsuk, Greenland
F.27-100	10119	J5-GBB	08/15/91	TAGB Air Bissau	Dori, Burkina Faso, Africa
F.27-100	10123	TC-TEZ	02/17/70	Turk Hava Yollari	Samsun, Turkey
F.27-100	10141	HB-AAI	09/13/64	Balair AG	Malaga, Spain
F.27-100	10142	PK-KFR	11/04/76	Bali Int'l Air	Sjamsuddin Noor, Banjar-Masin, Indonesia
F.27-100	10147	PI-C501	02/28/67	Philippine Airlines	Mactan Cebu, Philippines
F.27-200	10163	AP-ALM	08/06/70	Pakistan Int'l Airlines	Islamabad, Pakistan
F.27-100	10173	VT-DMC	10/19/88	Indian Airlines	Gauhatu, India
F.27-100	10174	VT-DMD	07/24/76	Indian Airlines	Bhubaneshwar, India
F.27-100	10175	VT-DME	08/12/72	Indian Airlines	Maqsoodpur, Delhi-Palam, India
F.27-200	10180	VH-FNH	03/17/65	Ansett ANA	Launceston, Australia
F.27-100	10182	TC-TAY	09/23/61	Turk Hava Yollari	Ankara, Turkey
F.27-200	10188	AP-ALX	12/12/71	Pakistan Int'l Airlines	Karachi-Zahedan, Pakistan
F.27-100	10191	PI-C503	10/12/62	Philippine Airlines	Paranaquerizal, Manilla, Philippines
F.27-200	10193	ST-AAR	07/02/85	Sudan Airways	location unknown

Type	MSN	Reg.	Date	Airline	Nearest major city
F.27-200	10194	ST-AAS	10/05/82	Sudan Airways	Merowe, Sudan
F.27-200	10205	CR-AIB	03/27/70	Mozambique Airlines	Gago Coutinho, Lourenco Marques, Mozambique
F.27-200	10206	PT-LCG	02/12/90	Brazil Central	Bauru, Brazil
F.27-200	10207	PH-FDW	08/25/89	Pakistan Int'l Airlines	Gilgit-Islamabad, Pakistan
F.27-200	10235	XY-ADK	03/25/78	Burma Airways	Rangoon, Burma
F.27-200	10250	AP-ATO	12/15/78	Pakistan Int'l Airlines	Karachi, Pakistan
F.27-200	10251	I-ATIP	04/16/72	Aero Trasporti Italiani	Ardinello di Amasend, Grasinone, Italy
F.27-200	10271	PH-SAB	02/02/66	Indian Airlines	Srinagar-Juma, India
F.27-400M	10273	ST-ADX	05/10/72	Sudan Airways	El Obeid, Sudan
F.27-200	10279	AP-ATT	10/08/65	Pakistan Int'l Airways	Naran, Kaghan Valley, Pakistan
F.27-400M	10282	ST-ADW	06/06/77	Sudan Airways	Fasher, Sudan
F.27-100	10285	PI-C527	07/06/67	Philippine Airlines	Bacolod, Negros Island, Philippines
F.27-400	10306	PK-PFB	04/27/67	Fokker	Malaybalay, Philippines
F.27-200	10308	S2-ABG	11/18/79	Bangladesh Biman	Savar Bazar, Dacca, Bangladesh
F.27-200	10309	VT-DVG	06/07/70	Indian Airlines	Agartala, India
F.27-100	10311	PI-C532	05/09/70	Philippines Airlines	Iligan Airport, Philippines
F.27-400	10313	XY-ADO	08/19/80	Burma Airways	Moulmein, Burma
F.27-600	10314	AP-AUS	12/08/72	Pakistan Int'l Airlines	Rawalpindi, Pakistan
F.27-600	10325	XY-AEK	02/03/89	Burma Airways	Rangoon, Burma
F.27-200	10330	AP-AUV	12/30/70	Pakistan Int'l Airlines	Shemshemagar, Pakistan
F.27-600	10335	AP-AUX	10/23/86	Pakistan Int'l Airlines	Peshawar, Pakistan
F.27-400	10336	VT-DWT	08/29/70	Indian Airlines	Kumbhirgram, Silchar, India
F.27-300	10356	TF-FIL	09/26/70	Icelandair	Myggenaes, Faroe Islands
F.27-200	10357	XY-ADP	06/21/87	Burma Airways	Hopong, Burma
F.27-600	10395	9Q-CLO	03/03/76	Air Zaire	locations unknown
F.27-600	10399	PK-MFD	05/09/91	Merpati Nusantara Airlines	Manado, Sulawezi, Indonesia
F.27-600	10402	9Q-CLP	02/08/80	Air Zaire	Kinshasa-Ndjili, Zaire
F.27-400M	10404	TC-72	03/16/75	Lineas Aereas del Estado	San Carlos de Bariloche, Argentina
F.27-600	10406	9Q-CLR	01/06/76	Air Zaire	Kisangani, Zaire
F.27-600	10410	PK-JFF	06/05/91	Sempati Air Transport	Gresik, Indonesia
F.27-400M	10411	TC-75	06/10/70	Lineas Aereas del Estado	Lima, Peru
F.27-400M	10416	TC-77	12/02/69	Lineas Aereas del Estado	Marambia, Antarctica
F.27-600	10422	PK-GFJ	09/07/74	Garuda Indonesia Airways	Tanjung Karang, Indonesia
F.27-500	10426	OY-APB	12/27/69	Maersk Air	Ronne, Denmark
F.27-500	10428	HL5212	01/23/71	Korean Air	Kansong, Korea
F.27-200	10439	CR-LLD	05/21/72	DTA Angola Airlines	Lobito, Angola
F.27-600	10443	OY-APE	05/26/88	Starair	Hannover, Germany
F.27-600	10452	XY-ADQ	06/16/88	Burma Airways	Putao, Burma
F.27-600	10453	S2-ABJ	08/05/84	Bangladesh Biman	Dhaka Airport, Bangladesh
F.27-500	10456	ZK-NFC	02/17/79	Air New Zealand	Auckland, New Zealand
F.27-600	10462	PK-GFP	09/26/72	Garuda Indonesian Airways	Jakarta, Indonesia
F.27-500	10463	9V-BCU	11/23/71	Malaysai-Singapore Airlines	Kota Kinabalu, Sabah
F.27-500	10473	C-GSFS	09/04/89	Securite Civile	La Grande Combe, Arles, France
F.27-600	10501	XY-ADS	10/12/85	Burma Airways	Putao, Burma
F.27-600RF	10557	6O-SAY	07/20/81	Somali Airlines	Balad, Sommali
F.27-600RF	10559	6O-SAZ	08/02/77	Somali Airlines	Borama, Somali
F.27-600	10562	D-AELB	02/24/90	FTG Air Service Flugcharter	Bergisch Gladbach, Germany
F.27-500	10570	F-BYAH	01/27/79	Air Rouergue	Rodez Airfield, France
F.27-600	10572	XY-ADY	10/03/78	Burma Airways	Mandalay, Burma
F.27-600	10573	TU-TIF	07/25/86	Air Ivoire	Tabou, Ivory Coast
F.27-600	10588	5A-DDV	06/06/89	Libyan Arab Airlines	Zella, Libya
F.27-400M	10649	9Q-CBH	12/13/92	Scibe Airlift Zaire	Goma, Zaire

Government Aircraft Factories Nomad

Type	MSN	Reg.	Date	Airline	Nearest major city
22	4	ZK-NOM	10/25/93	Hibiscus Air Service	Franz Joseph Glacier, New Zealand
22B	14	9M-ATZ	06/06/76	Sabah Air	Kota Kinabufu, Sabah
22B	28	PK-MAJ	07/23/79	Missionary Aviation Fellowship	West Irian, Indonesia
22B	39	P2-DNL	12/23/79	Douglas Airways	Menari, Papua Nugini
24A	44	N8071L	06/12/91	Tri Air	ditched off Great Inagua
24A	89	N418NE	05/04/90	Tar Heel Aviation	Wilmington, North Carolina

Grumman G-159 Gulfstream I

Type	MSN	Reg.	Date	Airline	Nearest major city
Gulfstream I	24	HK-3315	02/05/90	Helicol Columbia	Mt. El Saludo, Ibague, Columbia

Type	MSN	Reg.	Date	Airline	Nearest major city
Gulfstream I	59	HK-3316	05/02/90	Helicol Columbia	Los Garcones, Monteria, Columbia

Handley Page HPR.7 Herald
Type	MSN	Reg.	Date	Airline	Nearest major city
HPR.7-100	150	HK-718	02/11/73	La Urraca	Villavicencio, Columbia
HPR.7-101	152	HK-721	05/07/72	La Urraca	Valledupar, Columbia
HPR.7-201	157	B-2009	02/24/69	Far Eastern Air Transport	Kaohsiung, Taiwan
HPR.7-211	160	CF-NAF	03/17/65	Eastern Provincial Airways	Musquodoboit, Nova Scotia
HPR.7-203	168	I-TIVE	Apr 1971	Itavia	Rome, Italy
HPR.7-207	170	JY-ACQ	04/10/65	Alia Royal Jordanian Airlines	Damascus, Syria
HPR.7-401	178	HK-2701X	09/16/91	LACOL Columbia	Barranquilla, Columbia
HPR.7-401	180	HK-2702X	11/05/89	Aerosucre Columbia	Bogota, Columbia
HPR.7-214	190	PP-SDJ	11/03/67	SADIA	Marumbi Peak, Brazil

Handley Page HP.137 Jetstream
Type	MSN	Reg.	Date	Airline	Nearest major city
Mk.1	205	D-INAH	03/06/70	Bavaria Fluggesellschaft	Samedan, Switzerland
Mk.1	238	N11360	04/17/81	Air US	mid-air Cessna 206G N4862F, Loveland, Colorado

Hindustan Aeronautics Ltd. 748
Type	MSN	Reg.	Date	Airline	Nearest major city
2/224	506	VT-DUO	03/05/84	Indian Airlines	Hyderabad-Begumpet, India
2/224	511	VT-DXF	08/18/81	Indian Airlines	Mangalore, India
2/224	512	VT-DXG	12/09/71	Indian Airlines	Chinnamanur, Mandurai, India
2/224	514	VT-DXI	06/16/81	Indian Airlines	Tirupatui Airport, India
2/224	541	VT-EAU	03/15/73	Indian Airlines	Hyderabad, India

Lockheed 100 Hercules
Type	MSN	Reg.	Date	Airline	Nearest major city
100-30	4225	N14ST	05/23/74	Saturn Airways	Springfield, Illinois
100	4229	N760AL	12/24/68	Interior Airways	Prudhoe Bay, Alaska
100	4234	N102AK	10/27/74	Alaska Int'l Air	north of Fairbanks, Alaska
100-20	4303	S9-NAI	04/08/89	Transafrik	Luena, Angola
100-20	4333	N17ST	08/27/83	Transamerica Airlines	Dundo, Angola
100-20	4361	CF-PWX	11/21/76	Pacific Western Airlines	Kisangani, Zaire
100-20	4364	OB-R-1004	02/19/74	SATCO	Tarapoto, Peru
100-30	4701	HB-ILF	10/14/87	Zimex Aviation Ltd.,	Kuito, Angola
100-20	4830	D2-EAS	05/16/81	TAAG Angola Airlines	shot down by missile Mongua, Angola
100-30	4833	CP-1564	03/16/91	Transafrik	shot down by missile, Angola
100-30	4839	D2-EHD	01/22/93	Compania de Diamantes de Angola	Cachicala-Chiloango, Angola
100-30	5029	ET-AJL	09/17/91	Ethiopian Airlines	location unknown, Djibouti

Lockheed 188 Electra
Type	MSN	Reg.	Date	Airline	Nearest major city
188A	1007	RP-C1061	06/04/76	Air Manila Int'l	Agana, Guam
188A	1015	N6101A	02/03/59	American Airlines	East River, La Guardia, New York
188A	1018	HR-SAW	01/08/81	SASHA	Guatemala City, Guatemala
188A	1021	PI-C1060	01/09/72	Air Manilla Int'l	Pasay City, Philippines
188AF	1034	N5523	05/30/84	Zantop	Chalk Hill, Pennsylvania
188A	1044	N357Q	01/09/85	TPI Int'l Airways	Kansas City, Kansas
188A	1052	HC-AZY	09/12/88	TAME Ecuador	Lago Agrio, Ecuador
188C	1057	N121US	03/17/60	Northwest Orient Airlines	Tell City, Indiana
188AF	1059	N401FA	03/12/76	Great Northern Airlines	Lake Udrivik, Alaska
188A	1062	N5533	10/04/60	Eastern Air Lines	Boston, Massachusetts
188AF	1064	N400FA	12/11/74	Fairbanks Air Service	location unknown, Alaska
188AF	1076	N280F	07/06/77	Fleming Int'l Airways	St. Louis, Missouri
188A	1086	OB-R941	12/24/71	LANSA	Puerto Inca, Peru
188A	1087	HK-1976	07/10/75	Aerocondor Columbia	Bogota, Columbia
188A	1089	N283F	04/30/75	Zantop Int'l	Deadhorse, Alaska
188A	1090	N9705C	09/29/59	Braniff Airways	Buffalo, Texas
188A	1099	N9707C	05/03/68	Braniff Airways	Dawson, Texas
188C	1105	N126US	06/30/77	Coopertiva de Montecillos	into sea, Bocas del Toro, Panama
188A	1106	OB-R939	08/09/70	LANSA	Cuzco, Peru
188A	1115	HK-777	08/27/73	Aerocondor Condor	Bogota, Columbia
188A	1117	N6127A	09/14/60	American Airlines	La Guardia, New York.
188A	1121	N6130A	01/21/85	Galaxy Airlines	Reno, Nevada
188A	1127	N403GN	01/05/79	Great Northern Airlines	North Slope, Alaska
188CF	1134	HR-TNL	03/21/90	TAN Airlines	Las Mesitas, Honduras

Type	MSN	Reg.	Date	Airline	Nearest major city
188C	1136	N183H	04/22/66	American Flyers Airlines	Ardmore, Oklahoma
188C	1141	CF-PAB	10/29/74	Pan Artic Oils Ltd.	Rae Point, Melville Island
188C	1142	N137US	09/16/61	Northwest Orient Airlines	Chicago, Illinois
188C	2011	ZK-TEC	03/27/65	Tasman Empire Airways Ltd.	Whenaupai Airport, New Zealand
188CF	2012	N855U	08/24/70	Universal Airlines	Ogden, Utah
188C	2016	N859U	11/18/79	Transamerica Airlines	Granger, Utah
188C	2019	PH-LLM	06/12/61	KLM Royal Dutch Airlines	Cairo, Egypt
188C	2021	PK-GLB	02/16/67	Garuda Indonesian Airways	Menado, Indonesia

NAMC YS-11

Type	MSN	Reg.	Date	Airline	Nearest major city
YS-11-111	2023	JA8658	11/13/66	All Nippon Airways	Matsuyama Airport, Japan
YS-11A-211	2059	PP-SMI	04/12/72	VASP	Rio Del Janeiro, Brazil
YS-11A-202	2063	PP-CTG	10/18/72	Curzeiro do Sul del	Congonhas, Brazil
YS-11A-211	2068	PP-SMJ	10/23/73	VASP	Guanabara Bay, Brazil
YS011A-202	2082	N208PA	03/05/74	Pacific Southwest Airlines	Borrego Springs, California
YS-11A-301	2107	RP-C1419	07/17/77	Philippine Airlines	into sea off Mactan Airport, Philippines
YS-11A-219	2110	B-156	08/12/70	China Airlines	Yuan Shan, Taiwan
YS-11A-227	2134	JA8764	07/03/71	Japan Domestic Airlines	Mount Yokotsu-Dake, Hokkaido, Japan
YS-11A-300	2139	N128MP	03/15/89	Phoenix Airlines	Lafayette, Louisiana
YS-11A-213	2154	N906TC	01/13/87	Mid-Pacific Airlines	Lafayette, Louisiana
YS-11A-220500	2155	SX-BBQ	10/21/72	Olympic Airways	into sea, Athens, Greece
YS-11A-220500	2156	SX-BBR	11/23/76	Olympic Airways	Kozani, Greece

Nord 262

Type	MSN	Reg.	Date	Airline	Nearest major city
262E	2	F-BNGB	01/01/71	Rousseau Aviation	into Mediterranean Sea, Algiers, Algeria
262B-11	5	F-BLHT	11/12/73	Rousseau Aviation	Craon, Mayenne
262A-12	11	9Q-CJK	01/27/93	Trans Service Airlift	Kinshasa-Nkolo, Zaire
262A-12	19	7T-VSU	01/24/79	Air Algerie	Bechar-Adrar, Morocco
262A-33	41	N418SA	03/10/79	Swift Aire	into sea, Los Angeles, California
262A-27	47	N7886A	04/09/77	Altair Airlines	midair with Cessna 195 N4377N, Reading, Pa.
262A-36	48	N29824	02/12/79	Mohawk Airlines	Clarksburg, West Virginia

Short SC7

Type	MSN	Reg.	Date	Airline	Nearest major city
SC7-2-200	SH.1838	OH-SBB	11/01/89	RV Aviation	into sea off Aaland Islands, Finland
SC7-3-300	SH.1840	VN-PNI	09/01/72	Ansett Airlines of PNG	Mt Siluwe, Mendi, Papua New Guinea
SC7-2-200	SH.1841	N725R	04/14/72	Viking Int'l Air Freight	La Crosse, Wisconsin
SC7-3-200	SH.1850	N4917	08/25/77	Island Airways	Keahole Airort, Hawaii
SC7-3-200	SH.1858	N21CK	07/02/70	Jetco Aviation Int'l	Potomac River, Washington, D.C.
SC7-3-200	SH.1861	N123PA	10/22/70	Pan Alaska Airways	Fairbanks, Alaska
SC7-3-200	SH.1865	A40-SI	11/22/76	Gulf Air	into sea, Das Island, Abu Dhabi
SC7-3-100	SH.1892	G-OVAN	12/28/93	Peterborough Parachute Center	Ampuriabrava, Spain
SC7-3A-200	SH.1918	N20086	07/13/92	Circle Air Service	Bethel, Alaska
SC7-3-100	SH.1961	9M-PID	01/30/93	Pan-Malaysian Air Transport	Medan, Malaysia
SC7-3-100	SH.1963	9M-PIF	02/03/93	Pan-Malaysian Air Transport	Padang Marsirat, Langkawi Island, Malaysia
SC7-3-100	SH.1967	9M-AXM	07/30/93	Hornbill Airways	Miri, Sarawak
SC7-3-100	SH.1975	9M-AZB	09/03/91	Airtech Rajawali Udara	Long Seridan, Sarawak

Short SD-330

Type	MSN	Reg.	Date	Airline	Nearest major city
330-200	SH.3083	SX-BGE	08/03/89	Olympic Airways	Samos, Greece

Short SD-360

Type	MSN	Reg.	Date	Airline	Nearest major city
360-100	SH.3642	EI-BEM	01/31/86	Aer Lingus Commuter	East Midlands Airport, Derbyshire, England
360-300	SH.3719	EI-BTJ	12/13/87	Philippine Airlines	Mt. Gurain, Mindanao, Philippines

Vickers Vanguard

Type	MSN	Reg.	Date	Airline	Nearest major city
950	706	G-APEC	10/02/71	British European Airways	Aarsele, Ghent, Belgium
953	715	F-GEJF	01/29/88	Inter Cargo Service	Toulouse, France
952	730	F-GEJE	02/06/89	Inter Cargo Service	Marseilles, France
952	745	G-AXOP	04/10/73	Invicta Int'l Airlines	Jura Mtns., Hochwald, Switzerland

Vickers Viscount

Type	MSN	Reg.	Date	Airline	Nearest major city
701	4	G-ALWE	03/14/57	British European Airways	Wythenshawe, England
708	8	F-BGNK	12/12/56	Air France	Dannemois, France

Type	MSN	Reg.	Date	Airline	Nearest major city
701	25	G-AMOL	07/20/65	Cambrian Airways	Speke Airport, Liverpool, England
701	26	G-AMOM	01/21/56	British European Airways	Blackbushe Airport, Hampshire, England
708	39	F-BGNV	08/12/63	Air Inter	Lyon, France
720	44	VH-TVA	10/31/54	Trans Australia Airlines	Mangalore Aerodrome, Victoria, Australia
720	45	VH-RMQ	12/31/68	MacRobertson Miller Airlines	Port Hedland, Western Australia
720	46	VH-TVC	03/07/60	Ansett-ANA	Botany Bay, Sydney, Australia
724	50	F-BMCH	10/27/72	Air Inter	Mt. Pic du Picon, near Lyons, France
701C	63	G-ANHC	10/22/58	British European Airways	Anzio, Italy, mid-air with Italian Air Force F-86
701C	66	PP-SRR	09/04/64	VASP	Mt. Nova, Caledonia, Brazil
732	74	G-ANRR	12/02/58	Hunting Clan Air Transport	Frimley, Surrey, England
763D	82	YS-09C	03/05/59	TACA Int'l Airlines	Managua, Nicaragua
755D	91	CU-T603	11/01/58	Cubana	Nipe Bay, Cuba
749	94	YV-C-AMV	01/25/71	Linea Aeropostal Venezolana	Merida, Venezuela
749	95	YV-C-AMX	08/14/74	Linea Aeropostal Venezolana	Margarita Island, Venezuela
749	96	YV-C-AMZ	11/01/71	Linea Aeropostal Venezolana	Maracaibo, Venezuela
748D	102	VP-YNE	08/09/58	Central African Airways	Benghazi, Libya
745D	103	N7405	07/09/64	United Air Lines	Newport, Tennessee
745D	108	N7410	05/20/58	Capital Airlines	Brunswick, Maryland, mid-air with U.S. Air Force T-33
745D	128	N7430	11/23/62	United Air Lines	Ellicott City, Baltimore, Maryland
745D	135	N7437	04/06/58	Capital Airlines	Saginaw, Michigan
757	143	CF-TGY	10/03/59	Trans Canada Air Lines	Toronto, Canada
759D	149	TF-ISU	04/14/63	Icelandair	Nesoy Island, Norway
802	150	G-AOJA	10/23/57	British European Airways	Belfast-Nutts Corner, Northern Ireland
802	158	G-AOHI	01/19/73	British European Airways	Ben More, Perth, Australia
803	176	EI-AOF	06/22/67	Aer Lingus, Ashbourne	Dublin, Ireland
803	178	EI-AOM	03/24/68	Aer Lingus	Irish Sea Wexford Coast
764D	183	HC-BEM	12/29/77	Servicios Aereos Nacionales	Cuenca, Ecuador
764D	185	HC-BCL	09/04/77	Servicios Aereos Nacionales	Cuenca, Ecuador
768D	192	VT-DIO	09/11/63	Indian Airlines	Agra, India
745D	212	HK-2382	03/26/82	Aeropesca Columbia	Bogota, Columbia
745D	217	N7462	01/18/60	Capital Airlines	Charles City, Virginia
754D	244	OD-ADE	02/01/63	Middle East Airlines	Ankara, Turkey, mid-air with Turkish Air Force C-47
804	249	SP-LVA	08/20/65	LOT Polish Airlines	St. Trond, Belgium
802	254	G-AORC	04/28/58	British European Airways	Craigie, Scotland
806	260	PK-RVK	01/07/76	Mandala Airlines	Mando, Sulawesi, Indonesia
757	271	CF-THK	04/07/69	Air Canada	Sept-Iles, Quebec, Canada
745D	287	N7463	05/12/59	Capital Airlines	Chase, Maryland
768D	292	VT-DIX	08/09/71	Indian Airlines	Jaipur, India
782D	297	VP-WAS	09/03/78	Air Rhodesia	Whamira Hills near Lake Kariba, Rhodesia
818	317	ZS-CVA	03/13/67	South African Airways	East London, South Africa
785D	328	I-LAKE	03/28/64	Alitalia	Mt. Somma, Naples, Italy
785D	329	HC-AVP	04/23/79	SAETA, Quito	Cuenca, Ecuador
815	336	G-AVJA	03/20/69	British Midland Airways	Manchester, England
815	337	AP-AJE	08/14/59	Pakistan Int'l Airlines	Karachi, Pakistan
814	342	9G-ACL	06/10/78	West African Air Cargo	Sprigga Payne-Monrovia, Liberia
812	353	PK-IVS	08/26/80	Bouraq Indonesia Airlines	Jakarta, Indonesia
812	354	N243V	07/08/62	Continental Airlines	Amarillo, Texas
812	360	G-AVJZ	05/03/67	Channel Airways	Southend, Essex
838	372	SE-FOZ	01/15/77	Skyline Sweden	Kalvesta, Stockholm-Bromma, Sweden
785D	377	HC-ARS	08/15/76	SAETA	Quito-Cuenca, Ecuador
785D	378	I-LIZT	12/21/59	Alitalia	Rome-Ciampino, Italy
785D	379	6O-AAJ	05/06/70	Somali Airlines	Mogadishu, Somalia
785D	380	HK-1058	06/08/74	Aerolineas TAO	Monte San Isidro, Cucuta, Columbia
812	389	PK-RVW	10/07/83	Mandala Airlines	Semarang, Indonesia
798D	392	XA-SCM	02/27/92	Aero Eslava	Mexico City, Mexico
739A	394	G-AFTN	08/09/68	British Eagle Int'l Airlines	Langenbruck, West Germany
804	395	SP-LVB	12/19/62	LOT Polish Airlines	Okecie, near Warsaw, Poland
806	396	XW-TDN	03/00/75	Royal Air Lao	Phnom Penh, Cambodia
827	398	PP-SRD	05/15/73	VASP	Salvador, Brazil
827	399	PP-SRE	09/15/68	VASP	Sao Paulo, Brazil
827	401	PP-SRG	12/22/59	VASP	Rio de Janeiro, Brazil, mid-air with BAF AT-6

Type	MSN	Reg.	Date	Airline	Nearest major city
832	415	PK-RVN	05/01/81	Mandala Airlines	Semarang, Indonesia
832	416	VH-RMI	09/22/66	Ansett, Winton	Queensland, Australia
833	425	4X-AVC	10/26/69	Arkia-Israel Inland Airlines	Tel Aviv, Israel
739B	427	SY-AKW	09/29/60	United Arab Airlines	into sea near Elba
794D	429	TC-SEV	02/17/59	Turk Hava Yollari	Gatwick Airport, London, England
794D	432	TC-SET	02/02/69	Turk Hava Yollari	Ankara, Turkey
816	434	PK-RVU	07/24/92	Mandala Airlines	Liliboi, Jakarta, Indonesia
837	437	OE-LAF	09/26/60	Austrian Airlines	Moscow-Sheremetyevo, Russia
837	439	B-2029	07/31/75	Far Eastern Air Transport	Taipei, Taiwan
837	442	HK-1347	01/21/72	Lineas Aereas la Urraca	Funza, Columbia
828	444	JA8202	11/19/62	All Nippon Airways	Nagoya, Japan
828	448	PK-MVS	11/10/71	Merpati Nusantara Airlines	Padang, West Sumatra
843	454	PK-IVX	08/26/92	Bouraq Indonesian Airlines	Banjamasin, Indonesia

Jet Airliners

Airbus A.300

Type	MSN	Reg.	Date	Airline	Nearest major city
B2-320	122	OY-KAA	12/18/83	Malaysian Airline System	Kuala, Lumpur
B2-263	186	EP-IBU	07/03/88	Iran Air	into Persian Gulf, Bandar, Iran

Airbus A.320

Type	MSN	Reg.	Date	Airline	Nearest major city
-111	0009	F-GFKC	06/26/88	Air France	Habsheim, France
-111	0015	F-GGED	01/20/92	Air Inter	Strasbourg, France
-231	0079	VT-EPN	02/14/90	Indian Airlines	Bangalore, India

Boeing 707

Type	MSN	Reg.	Date	Airline	Nearest major city
707-121B	17586	N708PA	9/17/65	Pan Am	Chances Mtn., Antigua
707-121	17588	N709PA	12/08/63	Pan Am	Elkton, Maryland
707-124	17609	N70773	07/01/65	Continental	Kansas City Intl Airport, Missouri
707-328	17613	F-BHSA	07/27/61	Air France	Hamburg, Germany
707-329	17624	OO-SJB	02/15/61	Sabena	Brussels, Belgium
707-329	17627	OO-SJE	02/15/78	Sabena	Tenerife, Canary Islands
707-123	17629	N7502A	01/28/61	American Airlines	Montauk, New York
707-123	17633	N7506A	03/01/62	American Airlines	Jamaica Bay, New York
707-123	17641	N7514A	8/15/59	American Airlines	Calverton, New York
707-131	17671	N730JP	10/13/76	Lloyd Aero Boliviano	Santa Cruz, Bolivia
707-436	17706	G-APFE	03/05/66	BOAC	Mount Fuji, Japan
707-436	17712	G-APFK	03/17/77	British Airtours	Prestwick, Scotland
707-437	17722	VT-DJI	01/23/71	Air India	Bombay, India
707-437	17723	VT-DJJ	06/22/82	Air India	Bombay, India
707-441	17906	PP-VJB	11/27/62	Varig	Lima, Peru
720-023	18020	OD-AFT	01/01/76	Middle East Airlines	Al Qaysumah, Saudi Arabia
720-022	18044	N37777	04/22/76	US Global of Florida	Barrabquilla, Columbia
707-437	18055	VT-DMN	01/24/66	Air France	Mont Blanc, France
707-030B	18058	D-ABOK	12/04/61	Lufthansa	Ebersheim, Germany
707-328	18247	F-BHST	06/22/62	Air France	Pointe a Pitre, Guadeloupe
720-030B	18249	D-ABOP	07/15/64	Lufthansa	Ansbach, Germany
720-051B	18354	N724US	02/12/63	Northwest Orient	Everglades, Florida
720-040B	18379	AP-AMH	05/20/65	Pakistan Int'l Airlines	Cairo, Egypt
707-328B	18459	F-BHSZ	12/04/69	Air France	Caracas, Venezuela
707-330B	18463	D-ABOT	12/20/73	Lufthansa	New Dehli, India
707-321C	18579	G-BEBP	05/14/77	IAS Cargo Airlines	Lusaka, Zambia
707-351B	18584	CC-CCX	08/03/78	LAN Chile	Buenos-Aires, Argentina
707-373C	18707	HK-2401	12/14/83	TAMPA Columbia	Medellin, Columbia
707-331C	18712	N787TW	07/26/69	Trans World Airlines	Pomona, New Jersey
707-321C	18715	ST-ALX	03/24/92	Golden Star Air Cargo	Mount Hymittus, Greece
707-321C	18824	N799PA	12/26/68	Pan Am	Elmendorf Air Force Base, Alaska
707-321C	18826	CF-PWZ	01/02/73	Pacific Western Airlines	Edmonton, Canada
707-328C	18881	D2-TOV	07/21/88	TAAG Angolan Air Charter	Lagos, Nigeria
707-324C	18887	B-1834	09/11/79	China Airlines	into sea Taipei, Taiwan
707-338C	18955	SA-DJO	03/14/83	United African Airlines	Sheba, Libya
707-321B	18959	N417PA	07/22/73	Pan Am	Papeete, French Polynesia
707-331B	18978	N18701	12/22/75	TWA	Milan-Malpensa, Italy
707-327C	19106	PP-VLJ	06/09/73	Varig	Rio de Janeiro, Brazil
707-327C	19107	OD-AFX	07/23/79	Trans Mediterranean Airlines	Beirut, Lebanon

Type	MSN	Reg.	Date	Airline	Nearest major city
707-351C	19168	5N-AYJ	12/14/88	GAS Air Nigeria	Kom-Omran, near Luxor, Egypt
707-351C	19209	N144SP	04/13/87	Burlington Air Express	Kansas City, Missouri
707-329C	19211	OO-SJK	07/13/68	Sabena	Lagos, Nigeria
707-323C	19235	PP-VLU	01/30/79	Varig	Tokyo, Japan
707-321C	19268	N446PA	04/22/74	Pan Am	Singaraja, Bali
707-321B	19276	HK-2016	01/25/90	Avianca	NE JFK, New York
707-349C	19354	PT-TCS	03/21/89	Transbrazil	Sao Paulo-Guarulho, Brazil
707-321C	19363	HL7429	04/02/78	Korean Air Lines	Murmansk, Russia
707-321C	19368	N459PA	11/03/73	Pan Am	Boston, Massachusetts
707-321C	19371	N461PA	07/25/71	Pan Am	Manila, Philippines
707-321B	19376	N454PA	01/30/74	Pan Am	Pago Pago, Samoa
707-321C	19377	ST-SAC	12/04/90	Sudania Air Cargo	Nairobi, Kenya
720-047B	19439	N3166	03/31/71	Western Air Lines	Ontario, California
707-331B	19572	N7231T	02/08/89	Independent Air Inc.	Santa Maria, Azores
707-338C	19630	5X-UBC	10/17/88	Uganda Airlines	Fiumicino-Rome, Italy
707-321B	19696	N494PA	12/12/68	Pan Am	Caracas, Venezuela
707-344C	19705	ZS-EUW	04/20/68	South African Airways	Windhoek, South West Africa
707-373C	19715	HL-7412	08/02/76	Korean Air Lines	Teheran, Iran
707-328C	19724	F-BLCJ	03/05/68	Air France	Pointe a Pitre, Guadaloupe
707-360C	19736	ET-ACD	11/19/77	Ethiopian Airlines	Rome-Flumicino, Italy
707-379C	19822	PP-VJK	01/03/87	Varig	Point Alepe, Ivory Coast
707-345C	19841	PP-VJZ	07/11/73	Varig	Paris-Orly, France
707-366C	19845	SU-AOW	12/05/72	Egyptair	SE Cairo, Egypt
707-321B	20028	N320MJ	09/20/90	Omega Air	Marana, Arizona
707-331B	20063	N8734	09/08/74	TWA	Aegean Sea, Cephalonia, Greece
707-331C	20068	N15712	09/14/72	TWA	San Francisco Bay, California
707-366C	20342	SU-APE	10/17/82	Egyptair	Contrin Airport, Geneva, Switzerland
707-330C	20395	D-ABUY	07/26/79	Lufthansa	Petropolis, Rio de Janeiro, Brazil
707-340C	20487	AP-AVZ	12/15/71	Pakistan Int'l Airlines	Urumchi, China
707-3D3C	20494	JY-ADO	01/22/73	Alia Royal Jordanian Airlines	Kano, Nigeria
707-3B5C	20522	HL7406	11/29/87	Korean Air Lines	into Andaman Sea, Bangkok, Thailand
707-366C	20763	SU-AXA	12/25/76	Egyptair	Bangkok, Thailand
707-3K1C	21651	YR-ABD	01/10/91	Tarom	Bucharest, Romania

Boeing 727

Type	MSN	Reg.	Date	Airline	Nearest major city
727-22	18295	N68650	07/19/67	Piedmont	mid-air with Cessna 310 N31215, Hendersonville, North Carolina
727-22	18322	N7030U	04/07/65	United	Salt Lake City, Utah
727-22	18328	N7036U	08/16/65	United	Lake Michigan
727-81	18822	JA8302	02/04/66	All Nippon Airways	Tokyo Bay, Japan
727-22	18856	TI-LRC	05/23/88	LACSA	San Jose, Costa Rica
727-23	18901	N1996	11/08/65	American Airlines	Covington, Kentucky
727-21	18894	HK-2559	08/04/82	Aerotal Columbia	Simon Bolivar Airport, Columbia
727-21	18995	N317PA	11/15/66	Pan Am	East Berlin, Germany
727-21	18999	HK-1716	03/17/88	Avianca	Cucuta, Columbia
727-21	19035	HK-1803	11/27/89	Avianca	Bogota, Columbia
727-27C	19111	PT-TYS	04/12/80	Transbrazil	24 kms from Florianopolis Airport, Brazil
727-92C	19175	B-1018	02/16/68	Civil Air Transport	Taipei, Taiwan
727-64	19255	XA-SEJ	09/21/69	Mexicana	Lake Texcoco, Mexico City, Mexico
727-64	19256	XA-SEL	06/04/69	Mexicana	Monterey, Mexico
727-46	19279	G-BDAN	04/25/80	Dan-Air	Esperanza Forest, 20 kms S Tenerife Airport
727-193	19304	N2969G	09/04/71	Alaska Airlines	near Teardrop Lake, NE Juneau, Alaska
727-235	19457	N4737	07/09/82	Pan Am	New Orleans, Louisiana
727-224	19514	N88705	10/21/89	TAN Airlines	Tegucigalpa, Honduras
727-24C	19525	HK-1272	09/30/75	Avianca	Barranquilla, Columbia
727-214	19688	N533PS	09/25/78	Pacific Southwest Airlines	mid-air with Cessna 172 N7711G, San Diego, California
727-113C	19690	YA-FAR	01/05/69	Ariana Afghan Airways	London-Gatwick, England
727-116	19812	CC-CAQ	04/28/69	LAN Chile	Colina, Chile
727-86	19817	EP-IRD	01/21/80	Iran Air	Elburz mtns., Teheran, Iran
727-22C	19891	N7434U	01/18/69	United	into sea after take off, Los Angeles, California
727-247	20266	OB-1303	09/11/90	Air Malta	into sea off Newfoundland
727-251	20296	N274US	12/01/74	Northwest Orient Airlines	Bear Mountain Park, Stony Point, New York
727-231	20306	N54328	12/01/74	TWA	Upperville, Virginia

Type	MSN	Reg.	Date	Airline	Nearest major city
727-2E2/281	20436	JA8329	07/30/71	All Nippon Airways	Shizukuishi, Iwate, Japan, mid-air with JASDF F-86
727-225	20443	N8845E	06/24/75	Eastern Air Lines	on approach to JFK, New York, New York
727-232	20750	N473DA	08/31/88	Delta Air Lines	Dallas-Ft. Worth, Texas
727-2D3	20886	JY-ADU	03/14/79	Alia Royal Jordanian Airlines	Doha, Qatar
727-2H9	20930	TC-AKD	02/27/88	Talia Airways	Kyrenia, Turkey
727-2F2	20982	TC-JBH	09/19/76	Turk Hava Yollari	Mount Karatepe, Turkey
727-212	21347	PP-SRK	06/08/82	VASP	Sierra da Pacatuba mtns, Fortaleza, Brazil
727-256	21777	EC-DDU	02/19/85	Iberia	Mount Oiz, Spain
727-264	22414	XA-MEM	03/31/86	Mexicana	into mtns near Maravatio, Mexico
727-225	22556	N819EA	01/01/85	Eastern Air Lines	Mount Illimani, La Paz, Bolivia

Boeing 737

Type	MSN	Reg.	Date	Airline	Nearest major city
737-222	19069	N9031U	12/08/72	United Airlines	Chicago, Illinois
737-222	19556	N62AF	01/13/82	Air Florida	Washingtom, D.C.
737-222	19939	B-2603	08/21/81	Far Eastern Air Transport	Miao-Li, Taiwan
737-2A1	20096	PP-SME	01/28/86	VASP	Rio de Janeiro, Brazil
737-247	20131	N4527W	03/31/75	Western Airlines	Casper, Wyoming
737-281	20226	B-1870	02/16/86	China Airlines	into sea near Makung, Taiwan
737-2H6	20585	9M-MBD	12/04/77	Malaysian Airline System	Kuala Lumpur, Malaysia
737-229C	20914	OO-SDH	04/04/78	Sabena	Gosseiles, Belgium
737-241	21006	PP-VMK	09/03/89	Varig	Sao Jose do Xingu, Brazil
737-270C	21183	YI-AGJ	12/26/86	Iraqi Airways	Arar, Saudi Arabia
737-2A1C	21188	PP-SND	06/22/92	VASP	Cruzeiro do Sul, Brazil
737-2P6	21734	A40-BK	09/22/83	Gulf Air	Jebel Ali, Dubai
737-204	22059	HP-1205CMP	06/06/92	Copa Panama	Darien Pass, Tucuti, Panama
737-2P5	22267	HS-TBC	08/31/87	Thai Airways	into sea near Phuket, Thailand
737-2V2	22607	HC-BIG	07/11/83	TAME	Cuenca, Ecuador
737-230	22635	D-ABHD	01/02/88	Condor	Izmir, Turkey
737-2X6C	23121	N670MA	06/02/90	Markair	Unalakleet, Alaska
737-209	23795	B-180	10/26/89	China Airlines	Hualien, Taiwan
737-4YO	23867	G-OBME	01/08/89	British Midland Airways	Kegworth, Derbyshire, England
737-260	23914	ET-AJA	09/15/88	Ethiopian Airlines	Bahar Dar, Ethiopia

Boeing 747

Type	MSN	Reg.	Date	Airline	Nearest major city
747-121	19643	N736PA	03/27/77	Pan Am	Tenerife, Canary Islands
747-121SCD	19646	N739PA	12/21/88	Pan Am	Lockerbie, Scotland
747-130	19747	D-ABYB	11/20/74	Lufthansa	Nairobi, Kenya
747-237B	19959	VT-EBD	01/01/78	Air India	Bombay, India
747-206B	20400	PH-BUF	03/27/77	KLM-Royal Dutch Airlines	Tenerife, Canary Island
747-230B	20559	HL7442	09/01/83	Korean Air Lines	into Okhotsk Sea by Soviet missile, near Shakhalin Island
747SR-46	20783	JA8119	08/12/85	Japan Air Lines	NW Tokyo, Japan
747-283B	21381	HK-2910X	11/27/83	Avianca	Madrid, Spain
747-2B5B	21773	HL7445	11/19/80	Korean Air Lines	Seoul, South Korea
747-249F	21828	N807FT	02/19/89	Flying Tiger Line	Kuala Lumpur, Malaysia

Boeing 767

Type	MSN	Reg.	Date	Airline	Nearest major city
767-3Z9ER	24628	OE-LAV	05/26/91	Lauda Air	Bangkok, Thailand

BAe-146

Type	MSN	Reg.	Date	Airline	Nearest major city
BAe-146	E2027	N350PS	12/07/87	Pacific Southwest Airlines	San Luis Obispo, California

British Aircraft Corp. 1-11

Type	MSN	Reg.	Date	Airline	Nearest major city
1-11-200AB	004	G-ASHG	10/22/63	British Aircraft Corp.	Tisbury, Wiltshire, England
1-11-401AK	058	N711ST	02/09/75	Jet Travel	Lake Tahoe, California
1-11-070	070	N1553	08/06/66	Braniff Airways	Falls City, Nebraska
1-11-402AP	092	PI-C1131	09/12/69	Philippine Airlines	Manila, Philippines
1-11-402AP	094	5N-AOW	06/26/91	Okada Air	Sokoto, Nigeria
1-11-204AF	098	N1116J	06/23/67	Mohawk Airlines	Blossburg, Pennsylvania
1-11-424EU	130	YR-BCA	12/07/70	Tarom	Constanta Airport, Romania
1-11-420EL	155	LV-JGY	11/21/77	Austral Lineas Aereas	San Carlos de Bariloche, Argentina
1-11-515FB	207	D-ALAR	09/06/71	Paninternational	Hamburg, Germany
1-11-529FR	212	LV-LOX	05/07/81	Austral Lineas Aereas	Rio de la Plata, Brazil

Type	MSN	Reg.	Date	Airline	Nearest major city
1-11-527FK	248	RP-C1183	08/04/84	Philippine Airlines	Tacloban, Philippines

Convair 880

Type	MSN	Reg.	Date	Airline	Nearest major city
880-22-1	22-00-26	N820TW	09/13/65	TWA	Kansas City, Missouri
880-22-1	22-00-27	N821TW	11/21/67	TWA	Cincinnati, Ohio
880-22M-3	22-00-37M	VR-HFX	11/05/67	Cathay Pacific Airlines	Hong Kong
880-22M-3	22-00-45M	JA8030	08/26/66	Japan Domestic Airlines	Tokyo, Japan
880-22M-3	22-00-47M	N48060	08/21/76	Airtrust Singapore	Seletar Airport, Singapore
880-22M-3	22-00-49M	JA8028	06/25/69	Japan Air Lines	Moses Lake, Washington

Convair 990

Type	MSN	Reg.	Date	Airline	Nearest major city
990-30A-5	30-10-1	N711NA	04/12/73	NASA	mid-air with Navy P-3, Sunnyvale, California
990-30A-5	30-10-3	PK-GJA	05/28/68	Garuda Indonesian Airways	Bombay, India
990-30A-5	30-10-4	N7876	09/10/73	California Airmotive Corp.	Guam
990-30A-8	30-10-13	N5603	08/08/70	Modern Air Transport	Acapulco, Mexico
990-30A-6	30-10-15	HB-ICD	02/21/70	Swissair	Zurich, Switzerland
990-30A-5	30-10-25	EC-BZR	12/03/72	Spantax	Tenerife, Canary Islands
990-30A-5	30-10-32	EC-BNM		Spantax Arlanda	Stockholm, Sweden

DeHavilland DH.106 Comet

Type	MSN	Reg.	Date	Airline	Nearest major city
DH.106-1	06003	G-ALYP	01/10/54	BOAC	Elba, Italy
DH.106-1	06008	G-ALYV	05/02/53	BOAC	Calcutta, India
DH.106-1	06011	G-ALYY	04/08/54	BOAC	into Tyrrhenian Sea, Stromboli, Italy
DH.106-1	06012	G-ALYZ	10/26/52	BOAC	Rome-Ciampino, Italy
DH.106-1A	06014	CF-CUN	03/03/53	Canadian Pacific Airlines	Karachi, Pakistan
DH.106-4	6415	G-APDN	07/03/70	Dan-Air	Montesseny Mtns., Barcelona, Spain
DH.106-4	6430	LV-AHR	11/23/61	Aerolineas Argentinas	Sao Paulo, Brasil
DH.106-4C	6439	SU-ALC	01/02/71	United Arab Airlines	Ben Gashir, Libya
DH.106-4C	6441	SU-ALD	07/28/63	United Arab Airlines	into sea near Bombay, India
DH.106-4C	6444	SU-ALE	02/09/70	United Arab Airlines	Munich, Germany
DH.106-4B	6456	G-ARJM	12/21/61	British European Airways	Ankara, Turkey
DH.106-4C	6475	SU-ANI	01/14/70	United Arab Airlines	Addis Ababa, Ethiopia

Douglas DC-8

Type	MSN	Reg.	Date	Airline	Nearest major city
DC-8-32/33	45253	PP-PEA	03/04/67	Varig	Robertsfield, Monrovia
DC-8-33	45273	PP-PDT	08/20/62	Panair do Brasil	into sea near Rio de Janeiro, Brazil
DC-8-31	45277	N4903C	04/28/68	Capitol Int'l Airways	Atlantic City, New Jersey
DC-8-11	45290	N8013U	12/16/60	United Airlines	mid-air with TWA 1049 N6907C
DC-8-51	45409	N802E	03/30/67	Delta Airlines	New Orleans, Louisana
DC-8-21	45428	N8607	02/25/64	Eastern Air Lines	into Lake Pontchartrain, Louisana
DC-8-42/42	45598	OB-R1143	08/01/80	Aeronaves del Peru	Cerro Lilio, Mexico
DC-8-43	45611	CU-T1201	10/06/76	Cubana	into sea, Grantley Adams Int'l Airport, Barbados
DC-8-53	45615	PH-DCL	05/30/61	KLM - Royal Dutch Airlines	into sea en route to Santa Maria, Azores from Lisbon
DC-8-52	45617	EC-ARA	07/06/72	Aviaco	into sea Las Palmas, Canary Islands
DC-8-43	45625	I-DIWB	05/05/72	Alitalia	Mount Lunga, Palermo, Italy
DC-8-43	45631	I-DIWD	07/07/62	Alitalia	Bombay, India
DC-8-51	45633	XA-NUS	12/24/66	Aeronaves de Mexico	Lake Texcoco, Mexico
DC-8-51	45652	XA-PEI	08/13/66	Aeronaves de Mexico	Acapulco, Mexico
DC-8-52	45571	ZK-NZB	07/04/66	Air New Zealand	Auckland, New Zealand
DC-8-55F	45754	HC-BKN	09/18/84	AECA Aeroservicios Ecuatorianos	Quito, Ecuador
DC-8-55F	45818	PH-MBH	12/04/74	Garuda Indonesian Airways	Laxabana Hill, Colombo, Ceylon
DC-8-62	45822	LN-MOO	01/13/69	Scandinavian Airlines System	into sea Los Angeles, California
DC-8-55F	45859	5N-ARH	03/31/88	Arax Airlines	Cairo, Egypt
DC-8-54F	45860	9G-MKB	02/15/92	MK Air Cargo	Kano, Nigeria
DC-8-54F	45880	N8047U	12/18/77	United Airlines	Salt Lake City, Utah
DC-8-61	45889	JA8061	02/09/82	Japan Air Lines	Haneda, Japan
DC-8-63F	45923	N794AL	02/15/92	Burlington Air Express	Toledo Express Airport, Illinois
DC-8-63F	45929	N782AL	02/16/95	Air Transport International	Kansas City, Missouri
DC-8-63CF	45951	N4863T	09/08/70	Trans International Airlines	New York, New York
DC-8-61	45972	N8082U	12/28/78	United Airlines	Portland, Oregon
DC-8-61	45982	C-GMXQ	07/11/91	Nationair	Jeddah, Saudi Arabia
DC-8-63AF	46005	N785FT	07/27/70	Flying Tiger Line	Okinawa, Japan
DC-8-54F	46010	N8053U	01/11/83	United Airlines	Detroit, Michigan
DC-8-63F	46020	TF-FLA	11/15/78	Loftledir	Colombo, Ceylon

Type	MSN	Reg.	Date	Airline	Nearest major city
DC-8-63CF	46050	N8635	03/04/77	Overseas National Airways	Niamey, Nigeria
DC-8-62	46057	JA8040	11/28/72	Japan Air Lines	Moscow, Soviet Union
DC-8-63PF	46058	N950JW	12/12/85	Arrow Air	Gander, Newfoundland
DC-8-63CF	46060	N4909C	11/27/70	Capitol Int'l Airways	Anchorage, Alaska
DC-8-62	46107	N1809E	06/07/89	Surinam Airways	Paramaribo, Surinam
DC-8-63	46114	CF-TIW	07/05/70	Air Canada	Toronto, Canada
DC-8-63CF	46146	N802WA	09/08/73	Mount Dutton	Cold Bay, Alaska
DC-8-62AF	46148	JA8054	01/13/77	Japan Air Lines	Anchorage, Alaska
DC-8-62H	46152	JA8051	09/27/77	Japan Air Lines	Kuala Lumpur, Malaysia

Douglas DC-9

Type	MSN	Reg.	Date	Airline	Nearest major city
DC-9-14	45700	N3305L	05/30/72	Delta Airlines	Fort Worth, Texas
DC-9-15	45724	I-TIGI	06/27/80	Itavia	into sea Naples, Italy
DC-9-14	45726	N626X	11/15/87	Continental Airlines	Denver, Colorado
DC-9-14	45771	N8910E	02/09/79	Eastern Air Lines	Dade-Collier Airport, Florida
DC-9-15	45777	N1063T	03/09/67	Trans World Airlines	Urbana, Ohio
DC-9-14	45794	N9101	10/01/66	West Coast Airlines	Portland, Oregon
DC-9-14	45796	N9103	03/17/80	Texas Int'l Airlines	Baton Rouge, Louisana
DC-9-32	47032	N3323L	11/27/73	Delta Airlines	Chattanooga, Tennessee
DC-9-15	47034	N974Z	12/27/68	Ozark Air Lines	Sioux City, Iowa
DC-9-14	47056	YV-C-AVM	12/22/74	Avensa	Maturin, Venezuela
DC-9-31	47075	N975NE	07/31/73	Delta Airlines	into sea near Boston, Massachucetts
DC-9-32	47077	EC-BII	03/05/73	Iberia	Trauche, France
DC-9-15	47100	XA-SOC	06/20/73	Aeronaves de Mexico	Puerto Vallarta, Mexico
DC-9-33F	47192	N931F	03/18/89	Evergreen Int'l Airlines	Saginaw, Texas
DC-9-31	47211	N988VJ	09/09/69	Allegheny Airlines	Shelbyville, Indiana mid-air with PA-28
DC-9-32	47227	I-DIKQ	12/23/78	Alitalia	into sea Palermo, Sicily
DC-9-15F	47240	N565PC	02/17/91	Emery Worldwide Airlines	Cleveland, Ohio
DC-9-31	47245	N97S	11/14/70	Southern Airways	Huntington, West Virginia
DC-9-14	47309	N100ME	09/06/85	Midwest Express	Milwaukee, Wisconsin
DC-9-32	47356	XA-JED	08/31/86	Aeromexico	mid-air with Piper PA-28 Cerritos, California
DC-9-31	47393	N1335U	04/04/77	Southern Airways	New Hope, Georgia
DC-9-31	47400	N8984E	09/11/74	Eastern Airlines	Douglas Airport, North Carolina
DC-9-33F	47407	N935F	05/02/70	ALM-Antillean Airlines	St. Croix, Virgin Islands
DC-9-31	47441	N9345	06/06/71	Hughes Airwest	Duarte, California, mid-air with U.S. Marine Corps F-4 Phantom
DC-9-32	47457	YU-AJO	10/30/75	Inex Adria Airways	Prague, Czechoslovakia
DC-9-32	47482	YU-AHT	01/26/72	Jugoslovenski Aerotransport	Krussne Hory Mountain, Czechoslovakia
DC-9-32	47500	HI-177	02/15/70	Dominicana	into sea off Santo Domingo
DC-9-32	47621	XA-DEN	07/21/81	Aeromexico	Chihuahua, Mexico
DC-9-32	47622	XA-DEO	11/08/81	Aeromexico	Sierra de Guerreo, Mexico
DC-9-32	47641	I-ATHA	11/14/90	Alitalia	Zurich, Switzerland
DC-9-32	47649	YU-AJR	09/10/76	Inex Adria Airways	mid-air with Trident 3B G-AWZT Vrobec Village, Yugoslavia
DC-9-32	47667	I-ATJC	09/14/79	Aero Transporti	Italiani Sarroch, Caliari, Italy
DC-9-32	47720	YV-23C	03/05/91	Linea Aeropostal Venezolana	Torres Mtns, Venezuela
DC-9-32	47741	PK-GNQ	04/04/87	Garuda Indonesian Airways	Polonia, Indonesia
DC-9-82	48047	YU-ANA	12/01/81	Inex Adria Airways	Mont St. Pietro, Ajaccio, Corsica, France
DC-9-81	48050	N1003G	06/12/88	Austral Lineas Aereas	Posadas, Argentina
DC-9-82	48090	-----	08/16/87	Northwest Airlines	Detroit, Michigan
DC-9-81	53003	OY-KHO	12/27/91	Scandinavian Airlines System	Arlanda, Sweden

Douglas DC-10

Type	MSN	Reg.	Date	Airline	Nearest major city
DC-10-10	46510	N110AA	05/25/79	American Airlines	Chicago-O'Hare, Illinois
DC-10-10	46704	TC-JAV	03/03/74	Turk Hava Yollari	Senlis, France
DC-10-30	46852	N54629	09/19/89	Union de Transports	Aeriens Niger
DC-10-30	46925	EC-CBN	12/17/73	Iberia	Boston, Massachusetts
DC-10-30F	46960	HL7339	12/23/83	Korean Airlines	mid-air Piper PA-31 N35206 Anchorage, Alaska
DC-10-30	47887	HL7328	07/27/89	Korean Air	Tripoli, Libya

Fokker F.28

Type	MSN	Reg.	Date	Airline	Nearest major city
F.28-1000	11011	LN-SUY	12/23/72	Braathens SAFE	Fornebu, Norway
F.28-1000	11015	I-TIDE	01/01/74	Itavia	Turin, Italy
F.28-1000	11039	PK-GVC	09/24/75	Garuda Indonesian Airways	Palembang, Indonesia

Type	MSN	Reg.	Date	Airline	Nearest major city
F.28-1000	11055	PK-GVE	07/11/79	Garuda Indonesian Airways	Mount Sibayak, Indonesia
F.28-1000	11057	TC-JAO	01/26/74	Turk Hava Yollari	Izm, Turkey
F.28-1000	11058	TC-JAP	01/30/75	Turk Hava Yollari	into sea, Istanbul, Turkey
F.28-1000	11059	OB-R1020	10/25/88	Aero Peru	Juliaca, Peru
F.28-1000	11060	C-FONF	03/10/89	Air Ontario	Dryden, Ontario, Canada
F.28-1000	11071	TC-JAT	12/23/79	Turk Hava Yollari	Cucuk Koy, Turkey
F.28-1000	11086	LV-LOB	11/15/75	Aerolineas Argentinas	Concordia, Argentina
F.28-1000	11094	PK-GVP	03/06/79	Garuda Indonesian Airways	Mount Bromo, Probolinggo, East Java
F.28-4000	11141	PH-CHI	10/06/81	NLM Cityhopper	Moerdijk,
F.28-3000	11165	FAC-1140	03/28/85	Satena	San Vicente de Caguan, Columbia
F.28-4000	11223	HL7284	11/25/89	Korean Air	Seoul, South Korea
F.28-4000	11235	N485US	03/22/92	USAir	La Guardia, New York, NY
F.28-1000	11993	5N-ANA	03/02/78	Nigeria Airways	mid-air with MiG-21, Kano, Nigeria

Hawker Siddeley HS121 Trident

Type	MSN	Reg.	Date	Airline	Nearest major city
HS.121-1C	2109	G-ARPI	06/18/72	British Airways	Staines, England
HS.121-1C/1E	2118	9K-ACG	06/30/66		Kuwait Int'l Airport, Kuwait
HS.121-1C	2126	G-ARPY	06/03/66	Hawker Siddeley Aviation	Felthorpe, Norfolk, England
HS.121-2E	2170	B-266	04/26/82	Civil Aviation Admin of China	Mt Yangsu, China
HS.121	2172	B-274	03/14/79	Civil Aviation Admin of China	Canton, China
HS.121-	2320	G-AWZT	09/10/76	British Airways	mid-air with DC-9, YU-AJR Vrobec Village, Yugoslavia

Lockheed 1011 TriStar

Type	MSN	Reg.	Date	Airline	Nearest major city
1011-1	1011	N310EA	12/29/72	Eastern Airlines	Everglades, Miami, Florida
1011-1	1163	N726DA	08/02/85	Delta Airlines	Dallas-Fort Worth, Texas

Sud Aviation SE.210 Caravelle

Type	MSN	Reg.	Date	Airline	Nearest major city
SE.210-1	14	OY-KRB	09/27/61	Scandinavian Airlines System	Ankara, Turkey
SE.210-1	23	OD-AEM	04/17/64	Middle East Airlines	into sea off Dhahran, Saudi Arabia
SE.210-1A	25	HS-TGI	06/29/67	Thai Int'l	into sea off Hong Kong
SE.210-III	32	CN-CCV	04/01/70	Royal Air Maroc	Berrechid, Morocco
SE.210-III	40	HC-BAE	01/18/86	SAETA	Santa Elena, Guatemala
SE.210-III	68	F-BJTB	09/12/61	Air France	Rabat, Morocco
SE.210-VIN	69	OO-SRD	12/22/73	Royal Air Maroc	Mount Mellaline, Morocco
SE.210-III	122	B-1852	11/20/71	China Airlines	into sea off Penghu Island, Taiwan
SE.210-VIN	127	LV-HGY	07/03/63	Aerolineas Argentinas	Cordoba, Argentina
SE.210-10A/VIN	130	VT-DPP	02/15/66	Indian Airlines	New Dehli, India
SE.210-10A/VIN	134	VT-DSB	09/04/66	Indian Airlines	Bombay, India
SE.210-III	147	HB-ICV	09/04/63	Swissair	Duerrenaesch, Switzerland
SE.210-VIN	151	YU-AHD	09/11/73	Jugoslovenski Aerotransport	Titograd, Yugoslavia
SE.210-VIR	163	EC-ATV	01/07/72	Iberia	Atalayasa Mtns, San Jose, Costa Rica
SE.210-VIR	165	HK-1810	12/21/80	TAC Columbia	Riohacha Airport, Columbia
SE.210-10R	200	HB-ICK	12/18/77	SA de Transport Aerien	into sea off Funchal, Portugal
SE.210-10R	202	EC-BDD	11/04/67	Iberia	Black Down, Fernhurst, Sussex, England
SE.210-11R	219	HK-3288X	04/26/89	Aerosucre Columbia	Barranquilla, Columbia
SE.210-10R	225	EC-BIC	08/13/73	Aviaco	Coruna, Alvedro
SE.210-10R	228	EC-BID	03/05/73	Aviaco	Funchal, Portugal
SE.210-III	244	F-BOHB	09/11/68	Air France	into sea off Nice, France
SE.210-10B3	267	OY-STL	03/14/72	Sterling Airways	Dubai, United Arab Emerates

Vickers VC10

Type	MSN	Reg.	Date	Airline	Nearest major city
VC10-1101	804	5N-ABD	11/20/69	Nigeria Airways	Lagos, Nigeria

APPENDIX II

Piston Engine Airliner Crashes

This list is intended as a starting point for research purposes only. It does not list every aircraft of a particular type that has crashed. The authors/publishers make no guarantees that wreckage still exists at these sites. Wreck Chasers are cautioned to do their research and be prepared before setting out on an adventure.

Type	MSN	Reg.	Date	Airline	Nearest Major City
Airspeed Ambassador 2					
AS.57	5211	G-AMAD	07/03/68	BKS Air Transport	London-Heathrow, England
AS.57	5217	G-ALZU	02/06/58	British European Airways	Munich, West Germany
ATL.98 Carvair					
ATL.98	3/18339	G-ARSF	12/28/62	Channel Air Bridge	Rotterdam, The Netherlands
ATL.98	6/7480	CF-EPU	09/28/68	Eastern Provincial Airways	Twin Falls, Labrador
ATL.98	16/10485	HI-168	06/23/69	Dominicana	Miami, Florida
Avro Tudor					
688 Tudor I	1235	G-AGSU	08/23/47	Ministry of Supply & Aircraft Production	Woodford, Manchester
688 Tudor I	1253	G-AGRE	01/17/49	British South American Airways	into sea en route Bermuda to Jamaica
688 Tudor I	1255	G-AGRG	01/27/59	Air Charter	Brindisi, Italy
688 Tudor I	1256	G-AGRH	04/23/59	Air Charter	Mount Suphan, Turkey
688 Tudor I	1349	G-AHNP	01/30/48	British South American Airways	into Atlantic Azores to Bermuda
689 Tudor 2	1417	G-AkBY	03/12/50	Airflight Ltd.	Llandow, South Wales, England
Avro York					
C.1	1210	G-AGJD	02/01/49	BOAC	Castel Benito, Libya
C.1	MW136	CF-HMW	09/26/56	Maritime CentralnAirways	Chimo, Quebec, Canada
C.1	MW147	CF-HMZ	04/11/55	Associated Airways	Yellowknife, NWT, Canada
C.1	MW193	G-ANSY	02/25/56	Scottish Aviation	Luqa, Malta
C.1	MW231	EP-ADA	09/15/55	Persian Air Services	Basra, Iraq
C.1	MW248	——	07/04/48	Royal Air Force	mid-air with SAS DC-6 SE-BDA Middlessex, England
C.1	MW294	CF-HIQ	01/08/57	TransAir	Rankin Inlet, Hudson Bay, Canada
C.1	MW295	EP-ADE	07/15/59	Persian Air Services	Kuwait City, Kuwait
C.1	MW308	G-AMUL	04/30/56	Scottish Airlines	Stansted, Essex, England
C.1	MW321	G-AMUN	12/23/57	Scottish Airlines	Stansted, Essex, England
C.1	MW327	G-ANRC	09/22/54	Scottish Aviation	Stansted, Essex, England
C.1	MW332	G-AMUM	04/13/54	Scottish Aviation	Luqa, Malta
C.1	1218	OD-ACZ	03/15/63	Middle East Airlines	Karai, Iran
C.1	1219	G-AGNR	07/16/47	BOAC	AZ-Zubair, Basra, Iraq
C.1	1222	OD-ACO	05/24/61	Trans Mediterranean Airways	AZaiba, Muscat
C.1	1226	G-AGNY	06/26/54	Skyways Ltd.	Kyritz, Germany
C.1	1227	ZS-BRB	08/24/52	BOAC	Gatow, West Berlin
I	1300	G-AHEW	09/07/46	British South American Airways	Bathhurst, The Gambia
I	1301	G-AHEX	01/05/49	British South American Airways	Natal, Brazil
I	1303	G-AHEZ	04/13/47	British South American Airways	Dakar, Senegal
I	1304	G-AHFA	02/02/53	Lancashire Aircraft Corp.	into sea between Newfoundland to Jamaica
I	1307	OD-ADB	09/29/58	Middle East Airlines	into sea between Beirut to Rome
I	1311	OD-ADA	09/11/62	Middle East Airlines	Azaiba, Saudi Arabia
I	1316	G-AFHI	03/15/49	Skyways Ltd.	Gatow, West Berlin

Type	MSN	Reg.	Date	Airline	Nearest major city
I	1354	G-AMGL	03/11/52	Air Charter Ltd.	Hamburg, West Germany
I	1365	LV-XIG	12/23/46	Flota Aerea Mercante Argentina	Rio de Janeiro, Brazil
I	1366	LV-XIH	07/25/47	Flota Aerea MercanteArgentina	Buenos Aires, Argentina
C.1	PC4494	G-ALBX	06/19/49	Skyways Ltd.	Nuestadt, Germany

Boeing 377 Stratocruiser

Type	MSN	Reg.	Date	Airline	Nearest major city
377-10-26	15932	N1032V	03/26/55	Pan Am	into sea off Seattle, Washington
377-10-26	15933	N1033V	04/10/59	Pan Am	Juneau, Alaska
377-10-26	15939	N1039V	04/29/52	Pan Am	Carolina, Brazil
377-10-28	15943	G-ALSA	12/25/54	BOAC	Prestwick, Scotland
377-10-30	15954	N74608	04/02/56	Northwest	into sea off Seattle, Washington
377-10-29	15959	N90943	10/16/56	Pan Am	into sea between Honolulu and San Francisco
377-10-29	15960	N90944	11/09/57	Pan Am	into sea between Honolulu and San Francisco
377-10-34	15970	N31230	09/12/51	United Airlines	Redwood City, California

Bristol 170 Freighter

Type	MSN	Reg.	Date	Airline	Nearest major city
MK.IIB	12731	F-BHVB	11/04/58	Cle Air Transport	Le Touquet, France
Mk.IIA	12734	G-AHJB	07/04/46	Bristol Aeroplane Co.	into sea near Aracaju, Brazil
Mk.II	12738	F-BENF	07/28/50	Cle Air Transport	Tanezrouft, Tunbukta
Mk.II	12739	HC-SBU	08/06/49	Shell Co. of Ecuador	East Cordilleras, Ecuador
Mk.IIA	12742	G-AHJJ	03/21/50	Bristol Aeroplane Co.	Cowbridge, Glamorgan, Wales
Mk.21	12756	G-AICM	01/19/53	Silver City Airways	Templehof, West Berlin
Mk.21	12757	EC-ADI	05/09/57	Aviacion y Comercio	Barajas, Madrid, Spain
Mk.21E	12762	G-AICS	02/27/58	Manx Airlines	Horwich, Lancashire, England
Mk.21	12766	G-AIFF	00/00/49	Bristol Aeroplane Co.	into English Channel
Mk.21E	12772	HZ-AAC	10/08/57	Saudi Arabian Airlines	Taraif, Saudi Arabia
Mk.21	12779	VR-NAD	02/05/55	West African Airways Corp.	Calobar, Nigeria
Mk.21	12782	F-BECR	03/10/50	Ste Indochinoise de Transportes Aeriens	Siagon, South Vietnam
Mk.I	12788	F-BCJN	10/16/47	Societe Aerienne du Littoral	into sea Cartagena, Spain
Mk.XI	12792	SE-BNG	11/18/47	Trafik Turist Transportflyg	Revello, Italy
Mk.21	12797	EC-AEG	12/04/53	Aviacion y Comercio	Guaderrama Mtns. Spain
Mk.IIA	12802	F-BCJA	01/25/48	Ste IndoChinoise de Transports Aeriens	into sea en route Cyprus to Baghdad
Mk.21E	12807	VH-SJQ	05/10/75	Jetair Australia Party Ltd.	into Bass Strait, Victoria, Australia
Mk.21E	12809	F-VNAI	08/16/54	Air Vietnam	Pakse, Laos
Mk.21E	12812	F-BENG	04/11/48	Cie Air Transport, Los Barrios	Algecivon, Spain
Mk31	12828	ZK-AYH	11/21/57	Straits Air Freight Express	Christchurch, New Zealand
Mk.31	13073	G-AMWA	09/24/63	Bristish United Air Ferries	Guernsey, Channel Islands
Mk.31E	13076	EI-APM	06/12/67		Dublin, Eire
Mk.32	13078	CF-QWJ	03/00/77	Lambair Ltd.	Rankin Inlet, Canada
Mk.31M	13134	C-CYQY	06/21/88	Trans Provincial Airlines	Bronson Creek, B.C., Canada
MK.32	13260	G-ANWL	11/01/61	Silver City Airways	Les Prevosts, St Savior, Guernsey, Channel Islands

Britten-Norman Trislander

Type	MSN	Reg.	Date	Airline	Nearest major city
BN-2A Mk.III-2	1051	XA-JPE	05/05/89	Aerocozumel	Chichen Itza, Yucatan, Mexico

Canadair Argonaut

Type	MSN	Reg.	Date	Airline	Nearest major city
DC-4M1	103	CF-TEL	08/12/48	Trans Canada Airlines	Sydney, Nova Scotia, Canada
DC-4M2	128	CF-TFD	12/09/56	Trans Canada Airlines	Mount Siesse, BC, Canada
DC-4M2	137	I-ACOA	10/11/66	Hank Wharton	Garoua, North Cameroons
C-4-1	147	YV-C-LBV	07/06/62	Linea Expressa Bolivar	into sea off Caracas, Venezuela
C-4-1	148	CF-CPR	02/09/50	Canadian Pacific Air Lines	Tokyo, Japan
C-4-1	150	CF-TFW	04/08/54	Trans Canada Air Lines	mid-air with RCAF Harvard, Moose Jaw, Saskatchewan, Canada
C-4	151	G-ALHE	06/24/56	BOAC	Kano, Nigeria
C-4	153	G-ALHG	06/04/67	British Midland Airways	Stockport, Manchester
C-4	158	G-ALHL	09/21/53	BOAC	Idris, Tripoli, Libya
C-4	161	VP-KNY	04/11/62	East African Airways	Nairobi, Kenya

Type	MSN	Reg.	Date	Airline	Nearest major city
Convair 240					
240-1	3	N55VM	10/20/77	L&J Co.	Gillsburg, Mississippi
240-0	6	PP-CEV	01/15/65	Cruzeiro do Sul	Jabaguara, Brazil
240-1	22	N8405H	02/13/58	Western Air Lines	Garnet, California
240-0	25	PP-CEZ	05/09/52	Cruziero do Sul	Vitoria, Espirito Santo, Brazil
240-1	27	JA5088	05/29/65	Japan Domestic Airlines	Obihiro, Japan
240-1	37	N8407H	02/26/54	Western Air Lines	Wright, Wyoming
240-8	38	N7177B	09/04/78	Mission Air Lift	Clewiston, Florida
240-2	39	PP-VCK	09/22/58	Varig	Brasilia, Brazil
240-0	40	N94221	08/04/55	American Airlines	Fort Leonard Wood, Missouri
240-2	49	N90662	09/02/51	Pan Am	Kingston, Jamaica
240-7	52	AP-AEH	05/15/58	Pakistan Int'l Airlines	New Delhi, India
240-0	53	JA5098	02/27/64	Fuji Airlines	Oita, Japan
240-0	54	N94229	01/22/52	American Airlines	Elizabeth, New Jersey
240-2	55	JY-ACB	01/22/59	Air Jordan	Dabouk, Jordan
240-2	59	N90664	02/27/51	Mid-Continent Airlines	Tulsa, Oklahoma
240-4	61	HB-IRW	06/19/54	Swissair	into sea off Folkstone, England
240-0	66	N94234	03/20/55	American Airlines	Springfield, Missouri
240-2	71	XC-DOK	01/25/70	Carrier unknown	Poza Rica, Mexico
240-6	72	ZP-CDP	05/26/67	Lineas Aereas Paraguayas	Buenos Aires, Argentina
240-6	77	LV-ADQ	10/16/54	Aerolineas Argentinas	Capila del Senor, Argentina
240-4	78	D-BELU	07/31/60	Deutsche Flugdienst	Rimini, Italy
240-7	82	AP-AEG	03/14/53	Orient Airways	Kalahasahar, India
240-0	88	N94244	01/20/54	American Airlines	Buffalo, New York
240-2	90	N90670	08/15/58	Northeast Airlines	Nantucket, Massachusetts
240-2	103	PP-VCQ	12/22/62	Varig	Brasilia, Brazil
240-0	115	N300GR	05/24/80	Charles Clay	Spruce Creek, Florida
240-0	116	N94255	09/16/53	American Airlines	Albany, New York
240-0	117	N94256	02/19/61	Lyndon B Johnson	Johnson City, Texas
240-4	125	PH-TEI	05/25/53	KLM Royal Dutch Airlines	Amsterdam, The Netherlands
240-12	128	OO-AWO	12/19/53	Sabena	Zurich, Switzerland
240-0	139	N94266	06/22/49	American Airlines	Memphis, Tennessee
240-0	141	N777DC	04/01/78	Thrifty Threads	Ungula, Columbia
240-0	150	N94273	03/25/59	American Airlines	Chicago, Illinois
240-12	154	OO-AWQ	10/14/53	Sabena	Frankfurt, Germany
240-12	155	SP-LPB	04/11/58	LOT-Polish Airlinesn	Warsaw, Poland
240-12	157	N270L	01/17/70	Aspen Airways	Aspen, Colorado
240-13	158	N91238	01/14/52	Northeast Airlines	Flushing Bay, New York
240-13	161	N91241	08/11/49	Northeast Airlines	Portland, Maine
240-23	172	PK-GCB	02/27/62	Garuda Indonesian Airways	Palembang, Indonesia
240-23	175	PK-GCE	08/17/62	Garuda Indonesian Airways	Ambon, Indonesia
T-29B/VT-29B	261	XA-JOU	03/15/84	Aero Cozumel	Cancun, Mexico
T-29B/VT-29B	322	N155PA	06/19/85	Combs Airways	Trenton, Canada
Convair 340/440					
340-35	44	N90853	03/16/54	Continental Air Lines	Midland, Texas
340-32 (440)	45	N3414	11/29/66	Allegheny Airlines	Harrisburg, Pennsylvania
340-31	59	N73130	12/21/62	Frontier Airlines	Grand Island, Nebraska
340-32	71	N3422	07/17/55	Braniff Airways	Chicago, Illinois
340-38 (440)	95	N4809C	07/14/78	Charlie Inc.	South Caicos
340-41 (440)	112	YU-ADL	02/04/61	Jugoslovenski Aerotransport	Titograd, Yugoslavia
340-47 (440)	125	N8415H	07/23/65	Allegheny Airlines	Montoursville, Pennsylvania
340-38 (440)	138	N4819C	03/13/74	Sierra Pacific Airlines	White Mountain, Bishop, California
340-48 (440)	143	TG-ACA	04/27/77	Aviateca	Guatemala City, Guatemala
340-38	152	N4820C	03/20/68	Delta Air Lines	Evansville, Indiana
340-59	159	PP-CDW	05/03/63	Cruzeiro do sul	Sao Paulo, Brazil
340-58	162	YU-ADA	12/22/56	Jugoslovenski Aerotransport	Grub, Germany
340-68B	174	HZ-AAT	02/04/72	Saudi Atabian Airlines	Sanaa, Yemen
340-58	178	YU-ADC	10/10/55	Jugoslovenski Aerotransport	Vienna, Austria
340-62	191	PP-YRB	06/24/60	Real SA	Guanabara Bay, Rio De Janeiro, Brazil
340-62 (440)	195	SE-CCK	11/20/64	Linjeflyg	Angelholm, Sweden
340-68 (440)	210	N477KW	08/25/82	National Flight Services	Wolf Creek Pass, Colorado
340-68/C-131D	219	HZ-AAZ	07/08/68	Saudi Arabian Airlines	Dhahran, Saudi Arabia
440-35	319	CP-1489	04/20/85	Carga Aereo Transporta da Santa Rosa	

Type	MSN	Reg.	Date	Airline	Nearest major city
440-75	325	N21DR	01/17/82	Island Airlines Hawaii	Beni, Bolivia Pearl Harbor Channel, Oahu
440-11	327	XA-KEH	05/21/81	Aero Leon SA	Oaxaca, Mexico
440-11	335	HB-IMD	07/17/56	Swissair	Shannon, Eire
C-131E	342	N121CA	06/03/81	Command Aviation	Dillingham, Alaska
440-11	355	HB-IMF	02/10/67	Swissair	Regensberg, Switzerland
440-38	378	N102US	00/00/80	Farm-Kem	Colombia
440-62	388	EC-ATH	03/31/65	Iberia	into sea off Tangier, Morocco
440-75	395	SE-BSU	11/01/69	Linjeflyg	Arlanda-Stockholm, Sweden
440-11	414	CP-1358	01/11/80	Carga Aero Transportadora	Camiare, Beni, Bolivia
440-86	415	CP-1078	03/26/76	San Francisco Servicio Aereos	into Pacific, near Chile
440-86	416	N9302	05/12/78	Argosy Air Lines	Skippingport, Pennsylvania
440-86	425	CP-924	09/06/78	North East Bolivian Airways	El Desengano, Yacuma, Bolivia
440-62	443	EC-ATB	11/12/62	Iberia	Carmona, Seville, Spain
440-88	448	N985	11/30/76	OMCO Petroleum Co.	Cairo, Egypt
440-62	456	PP-AQE	09/18/57	REAL SA	Montevideo, Uruguay
440-0	464	D-ACAT	01/28/66	Lufthansa	Bremen, Germany
440-59	493	PP-CEP	06/16/58	Cruziero do Sul	Curitaba, Parana, Brazil
de Havilland DH-114 Heron					
1B	14001	G-APJS	02/19/58	Gulf Aviation	Mount Saraceno, Italy
1B	14004	PP-SLG	03/27/53	Transportes Aereos Salvador	Itapebi, Bahia, Brazil
2 (2D)	14007	G-AMTS	07/10/61	Metropolitan Air Movements	Biggin Hill, Kent, England
1B	14009	HR-SAN	05/26/70	Aeroservice S de RL	Tegucigalpa, Toncontin
1B	14010	F-BGOI	04/18/55	Union Aeromaritime de Transport	Kupe Mountain
1B	14017	G-AMYU	08/15/58	Jersey Airlines	Guernsey, Channel Islands
1B	14026	JA6159	08/17/63	Fuji Airlines	Mount Hachijo, Fuji, Japan
1B	14028	JA6158	02/23/62	Toa Airways	Mount Ohira, Hofu-shi, Yamaguchi, Japan
1B	14033	VQ-FAL	12/11/65	Fiji Airways	Tavenni
1B	14034	TN-ABA	07/11/69	Cogair	Ruddervoorde, Belgium
1B	14040	F-OCED	01/27/68	Air Comores	Moroni, Comores
2B	14052	VP-BAO	08/23/59	LIAT	St Kitts
2B	14056	DQ-FEF	12/27/86	Sunflower Airlines	Nadi, Fiji
2	14067	VH-CLS	10/23/75	Connair	Cairns, Australia
2B	14085	N554PR	06/24/72	Prinair	Mercedita Airport
1B	14099	G-AOFY	09/28/57	British European Airways	Port Ellen, Isle of Islay
2D	14109	N19D	03/29/72	Air Pacific	location unknown
2	14111	N998SA	12/18/78	Fisher Bros Aviation	Cleveland, Ohio
2D	14113	EC-ANJ	04/14/58	Aviaco	Barcelona, Spain
2D	14117	EC-ANZ	11/15/57	Aviaco	Palma deMajorca, Spain
2D	14125	N563PR	03/05/69	Prinair	San Juan, Puerto Rico
2D	14127	LN-NPH	08/04/68	Norflyselskap	Bodo, Norway
2D	14132	CR-TAI	01/26/60	Transportes Aereos de Timor	into sea off Timor
Douglas DC-4					
C-54-DO	3060	NC88839	04/18/57	Capital Airlines	Pittsburgh, Pennsylvania
C-54A-DO	3065	7T-VAU	04/11/67	Air Algerie	Tamanrassett, Algeria
C-54A-DO	3076	N79998	08/15/49	Transocean Air Lines	into sea off Lunga Point, Ireland
C-54A-DO	3079	F-BDRI	07/18/51	Transports Aeriens Intercontinentaux	Tananarive, Madagascar
C-54A-DO	3084	F-BELB	12/08/50	Transports Aeriens Intercontinentaux	Bangui, Central Africa
C-54A-DO	3094	VT-DIA	09/08/58	Indian Airlines	Madras, India
C-54A-DO	3098	XW-TDE	02/11/72	Royal Air Lao	between Saigon and Vientiane, Laos
C-54A-DO	3099	OO-SBL	04/22/60	Sobelair	Bunia, Belgian Congo
C-54A-DO	3106	PI-C100	01/11/47	Far Eastern Air Transport	Lasaq
C-54-DO	3112	N88842	06/13/47	Pennsylvania Central Airlines	Lookout Rock, West Virginia
C-4-DO	3114	41-32939	01/15/43	Trans Continental & Western	Paramaribo, Surinam
C-54-DO	3123	N88852	11/17/55	R.R. Kerkorian	Seattle, Washington
C-54A-1-DO	7445	LV-ABI	09/27/49	Flota Aerea Mercante Argentina	Castilla, Argentina
C-54A-1-DO	7449	N68736	05/13/57	U.S. Overseas Airlines	Narssarssuaq, Greenland
C-54A-1-DO	7459	D-ALAF	11/03/57	Karl Herfurtner	Dusseldorf, Germany
C-54A-1-DO	7462	OB-R247	12/30/76	Faucett	Andes Mountains, Peru

Type	MSN	Reg.	Date	Airline	Nearest major city
C-54A-1-DO	7464	G-AJPL	02/04/49	Skyways Ltd.	Castel Benito, Lybia
C-54A-5-DO	7467	PP-AXS	11/04/57	REAL,	into sea off Praia de Balaia, Brazil
C-54A-5-DO	7468	LV-ABQ	06/17/53	Aerolineas Argentinas	Cordova, Argentina
C-54A-5-DO	7473	TJ-ABC	06/13/61	Air Cameroon,	Douala, Cameroon
C-54A-5-DO	7479	LV-AFG	01/13/48	Flota Aerea Mercante Argentina	Ponta Galeria, Italy
C-54A-1-DC	10270	N05425	06/23/50	Northwest Airlines	Benton Harbor, Lake Michigan
C-54A-1-DC	10274	CP-717	05/29/64	Transportes Aereos	Benianos, Trinidad, Bolivia
C-54A-1-DC	10277	OB-PAZ-228	10/02/55	Faucett	Pico Oiriruma-Vinac, Peru
C-54A-1-DC	10279	G-APYK	06/03/67	Air Ferry	Mount Canigou, France
C-54A-5-DC	10280	HK-172	01/15/76	Taxi Aereo	Bogota, Columbia
C-54A-5-DC	10284	OB-R-148	12/08/67	Faucett	Carpich Huanuco Mountain, Peru
R5D-1	10290	F-BFGR	02/07/53	Compagnie Autrex-Lopez-Loreta-Lebreton	Eysines, France
R5D-1	10294	CP-747	05/29/64	Transportes Aereos Benianos	location unknown
C-54A010-DC	10310	VR-HEU	07/23/54	Cathay Pacific	into sea off Hainan Island, China
C-54A-10-DC	10315	N4726V	03/28/64	Facilities Management Corp.	into sea between San Francisco and Hawaii
C-54A-10-DC	10317	N88920	10/26/47	Pan Am	Annette Island, Alaska
R5D-1	10319	CU-T-397	12/06/52	Cubana	Bermuda
R5D-1	10320	N48762	10/28/60	Northwest Airlines	Missoula, Montana
R5D-1	10327	CF-CPC	07/21/51	Canadian Pacific Air Lines	en route Sitka-Yakutat, Canada
C-54A-10-DC	10328	LV-JPG	02/19/71	Aero Palas	Cervo Huaycas, Peru
C-54A-10-DC	10331	N30061	11/14/61	Zantop	Cincinnati Airport, Kentucky
C-54A-10-DC	10339	N90439	02/05/50	El Al	Lydda, Israel
R5D-1	10341	F-BFCQ	04/04/54	Compagnie Autrex-Lopez-Loreta-Lebreton	Gian Lan, Vietnam
R5D-1	10342	N91068	09/12/46	Pennsylvania Central Air Lines	Washington, D.C.
C-54A-15-DC	10346	F-BFCP	05/28/69	Air France	Paris, France
C-54A-15-DC	10347	N5519V	03/22/61	Seven Seas Air Lines	Nagpur, India
C-54A-15-DC	10356	N3821	03/09/69	Continental Air Transport	into Atlantic
C-54A-15-DC	10359	G-ASOG	01/21/67	Air Ferry	Frankfurt, West Germany
C-54A-15-DC	10360	CF-ILI	11/04/59	Wheeler Airlines	St. Cleophas, Quebec, Canada
C-54A-15-DC	10376	VT-DIC	05/07/62	Indian Airlines	Haveri, India
C-54A-15-DC	10384	EP-ADK	08/04/61	Iran Air	location unknown
C-54A-15-DC	10387	HP-382	07/10/63	Aerovias Panama Airways	into sea off South Caicos Island
C-54A-15-DC	10388	N30065	05/29/46	United Air Lines	Chicago, Illinois
C-54A-15-DC	10391	F-BELO	11/28/49	Air France	St Just Chalignin, France
C-54A-15-DC	10403	HK-1309	06/30/75	Taxi Aereo El Canado	Saravena Arauca, Columbia
C-54A-15-DC	10406	N88785	10/27/48	Northwest Airlines	Edmonton, Alberta, Canada
C-54A-15-DC	10410	N90433	09/24/55	Flying Tigers	into sea off Honolulu, Hawaii
C-54A-15-DC	10418	HK-135	06/23/59	Avianca	Cerro Baco Mountain, Peru
C-54A-15-DC	10424	OD-AEB	12/12/63	Trans Mediterranean Airways	Koh Mountains, Kabul, Afghanistan
R5D-2	10431	N98AS	03/28/81	Caribbean Air Cargo	into sea off St. Croix, Virgin Islands
C-54B-1-DC	10439	C-114	02/15/47	Avianca	Sabana Mountains, Columbia
C-54B-1-DC	10441	9Q-CAM	11/26/77	African Lux	Tete Province, Mozambique
C-54B-1-DC	10445	N90427	09/28/68	Pan African Airlines	Port Harcourt, Nigeria
C-54B-1-DC	10450	N30050	12/24/46	United Air Lines	Los Angeles, California
C-54B-1-DC	10452	9Q-CBG	08/23/88	Kinair Cargo	River Zaire, Zaire
C-54B-1-DC	10469	HK-130	01/28/62	Avianca	Cucuta, Columbia
C-54B-1-DC	10471	N30051	10/08/46	United Air Lines	Cheyenne, Wyoming
C-54B-1-DC	10486	N63396	09/24/59	Reeve Aleutian Airways	Adak, Alaska
C-54B-1-DC	10493	XA-FOW	06/25/46	Mexicana	Mexico City, Mexico
C-54B-1-DC	10498	F-BFVO	11/06/52	Union Aeromaritime de Transport	Lake Chad, Chad
C-54B-1-DC	10503	N88899	04/11/52	Pan Am	into sea off San Juan, Puerto Rico
C-54B-1-DC	10504	N88900	09/12/59	Pan Am	Tegucigalpa, Honduras
C-54B-1-DC	10506	PP-BTQ	04/04/64	Paraense	Belem, Brazil
C-54B-1-DC	10507	HK-728	07/24/74	Aeronorte Columbia	Bogota, Columbia
C-54B-1-DC	10510	CP-609	02/05/60	Lloyd Aero Boliviano	Laguna de Juana-Costa, Bolivia
C-54B-1-DC	10512	4X-ADN	11/24/51	El Al	Zurich, Switzerland
C-54B-1-DC	10513	N9982H	05/24/69	Pan African Airlines	Port Harcourt, Nigeria
C-54B-1-DC	10518	CF-EDN	11/13/50	Curtiss Reid Flying Svce.	Grenoble, France
C-54B-1-DC	10530	D-ABEB	06/17/61	Cont. Deutsche Luftreederei	Kano, Nigeria
C-54B-1-DC	10544	PP-LEQ	08/12/58	Loide Aereo Naciona	Belem, Brazil
R5D-3	10559	CP-1206	03/24/84	Frigorifico Reyes	Rurrenabaque, Bolivia

Type	MSN	Reg.	Date	Airline	Nearest major city
R5D-3	10563	N8060C	11/19/79	Tiburon Aircraft Inc.	McCormick, South Carolina
R5D-3Z	10576	CP-1404	06/08/83	Frigorifico Reyes	Trinidad, Bolivia
C-54D-1-DC	10588	XV-NUH	09/24/72	Air Vietnam	Ben Cat, South Vietnam
C-54D-1-DC	10591	OB-R-776	01/14/70	Faucett	Mount Pumacona, Lima, Peru
C-54D-10-DC	10761	CP-1352	01/12/79	Trak Airways	San Borja, Bolivia
C-54D-10-DC	10832	N79992	11/17/51	Overseas Nat'l Airways	mid-air with DC-4 N4002B
C-54D-10-DC	10853	FAC-1106	12/18/79	Satena	Cerro Toledo Mountain, Columbia
C-54B-1-DO	18325	HK-730	01/15/66	Avianca	into sea off Cartagena, Columbia
C-54B-5-DO	18330	N95412	06/01/47	Northwest Airlines	Chicago, Illinois
C-54B-5-DO	18336	PP-LEM	02/01/58	Loide Aereo Nacional	Rio de Janeiro, Brazil
C-54B-5-DO	18348	N8342C	12/21/48	Civil Air Transport	Basalt Island, Hong Kong
C-54B-10-DO	18350	N86574	01/07/53	Flying Tigers	Issaquah, Washington
R5D-2	18365	N88727	11/01/49	Eastern Air Lines	mid-air with P-38 NX26927, Wash., D.C.
C-54B-10-DO	18367	YA-BAG	11/21/59	Ariana Afghan Airlines	Beirut, Lebanon
C-54B-15-DO	18374	CF-MCF	08/11/57	Maritime Central Airways	Issoudun, Quebec
C-54B-15-DO	18379	N384	03/10/64	Slick Airways	Boston, Massachusetts
R5D-2	18380	N88814	05/30/47	Eastern Air Lines	Havre de Grace, Maryland
C-54B-15-DO	18384	OO-DEP	11/29/64	Belgian Int'l Air Svcs	Leopoldville, Belgian Congo
C-54B-15-DO	18386	N44567	02/06/47	Aerovias Cuba Int'l	Pedro Bernados, Spain
C-54B-15-DO	18389	N30062	10/06/55	United Air Lines	Laramie, Wyoming
C-54B-15-DO	18390	N37478	03/10/48	Delta Air Lines	Chicago, Illinois
R5D-2	18396	N88729	10/11/46	Eastern Air Lines	Alexandra, Virginia
C-54B-15-DO	18397	N88939	01/00/85	BIF Air Inc	South Caicos, Caicos Islands
C-54D-15-DC	22174	XV-NUI	03/19/73	Air Vietnam	Ban e Thout, South Vietnam
C-54B-20-DO	27229	N88911	09/20/47	Pan Am	New York, New York
C-54B-20-DO	27239	N90449	03/02/57	Alaska Airlines	Blyn, Washington
C-54B-20-DO	27240	TF-RVC	09/14/50	Loftleidir	Catna Jokull Glacier, Iceland
C-54B-20-DO	27242	A2-ZER	03/04/74	Wenela	Francistown, Botswana
C-54E-5-DO	27279	N45342	01/19/52	Northwest Airlines	Sandspit, British Columbia, Canada
C-54E-5-DO	27290	N90904	10/03/46	American Overseas Airlines	Stephenville, Newfoundland
R5D-4R	27368	N96361	12/23/86	Central Air Service	Arlington, Washington
C-54E-15-DO	27374	F-RAFA	01/21/82	Aeronavale	Noumea, Tahiti
C-54D-10-DO	36061	N3373F	07/14/81	Aero Union	Kenai, Alaska
C-54G-10-DO	36072	N300JT	06/29/83	Pacific Air Express	Kahului, Hawaii
C-54D-10-DO	36076	N88942	03/20/53	Transocean Air Lines	Alvarado, California
DC-4-1009	42910	VH-ANA	06/26/50	Australian National Airways	York, Western Australia
DC-4-1009	42912	F-BBDB	01/22/50	Air France	Paris, France
DC-4-1009	42918	OD-AEC	07/09/62	Trans Mediterranean Airways	Brindisi, Italy
DC-4-1009	42930	SE-BBG	10/20/47	AB Aerotransport	Athens, Greece
DC-4-1009	42932	OO-CBE	05/13/46	Sabena	Magazini, Congo
DC-4-1009	42935	F-BBDC	04/10/48	Air France	Kano, Nigeria
DC-4-1009	42937	F-BBDE	06/12/50	Air France	Bahrain, Saudi Arabia
DC-4-1009	42950	VH-AND	10/16/52	Australian National Airlines	Sydney, Australia
DC-4-1009	42982	JA6011	09/30/57	Japan Air Lines	Itami, Osaka, Japan
DC-4-1009	42986	OO-CBG	09/18/46	Sabena	Gander, Newfoundland
DC-4-1009	42989	F-BBDL	07/12/48	Air France	Paris, France
DC-4-1009	42990	F-BBDM	05/14/60	Air France	Bahrain, Saudi Arabia
DC-4-1009	42991	SR-MAD	07/19/67	Air Madagascar	Tanararive, Madagascar
DC-4-1009	42992	F-BBDO	02/03/51	Air France	Mount Cameroun, Nigeria
DC-4-1009	43065	VH-TAA	05/24/61	Trans Australia Airlines	Buiwer Island, Brisbane, Australia
DC-4-1009	43072	HB-ILA	05/15/60	Ballair AG	Jebbel Marra, Sudan
DC-4-1009	43073	HB-ILE	12/13/50	Swissair	Sydney, Nova Scotia
DC-4-1009	43095	9Q-CBH	04/07/74	Air Zaire	Gemena, Zaire
DC-4-1009	43097	YK-AAR	09/01/60	Syrian Airways	Kinshasa, Congo
DC-4-1009	43098	HB-ILO	12/14/51	Swissair	Schiphol, Amsterdam

Douglas DC-6

Type	MSN	Reg.	Date	Airline	Nearest major city
XC-112A	35325	N901MA	02/08/76	Mercer Airlines	Van Nuys, California
DC-6	42871	N37506	06/17/48	United Air Lines	Mount Carmel, Pennsylvania
DC-6	42875	N37510	10/24/47	United Airlines	Bryce Canyon, Utah
DC-6A	42901	N90806	07/12/53	Transocean Air Lines	Wake Island
DC-6	42902	PI-C294	01/14/54	Philippine Air Lines	Rome, Italy
DC-6	43001	N37512	04/04/55	United Air Lines	Ronkonkoma, New York
DC-6	43031	LV-ADS	09/07/60	Aerolineas Argentinas	Salto, Uruguay

Type	MSN	Reg.	Date	Airline	Nearest major city
DC-6	43034	LV-ADV	06/10/58	Aerolineas Argentineas	Rio de Janeiro, Brazil
DC-6	43043	CP-926	06/17/71	Savco	Arica, Chile
DC-6	43055	N90891	02/11/52	National Airlines	Newark, New Jersey
DC-6	43057	N90893	02/14/53	National Airlines	Mobile, Alabama
DC-6	43059	XA-LAU	01/30/67	Mexicana	Merida, Mexico
DC-6	43112	PT-TKW	05/01/48	KLM-Royal Dutch Airlines	Amsterdam, The Netherands
DC-6	43114	PT-TPJ	03/22/52		Frankfurt, Germany
DC-6	43119	SE-BDA	07/04/48	Scandinavian Airlines	mid-air with Avro York MW248, Northwood, Middlesex, England
DC-6	43125	VH-BPE	10/29/53	BCPA	Woodside, California
DC-6	43126	XW-PEH	02/01/72	Penas	Tegal, Java
DC-6	43133	XA-NAH	06/29/67	Saesa	Mazatlan, Mexico
DC-6	43136	LV-ADW	07/19/61	Aerolineas Argentinas	Azul, Argentina
DC-6	43144	N37543	06/30/51	United Air Lines	Fort Collins, Colorado
DC-6	43212	XA-JOS	09/01/51	Mexicana	Mexico City, Mexico
DC-6	43215	I-LUCK	12/23/51	Linee Aeree Italiane	Milan, Italy
DC-6B	43260	N37550	08/24/51	United Air Lines	DeCoto (Union City), California
DC-6B	43270	HI-146	10/19/71	LADECO	Mendoza, Argentina
DC-6B	43273	CP-698	09/26/69	Lloyd Aereo Boliviano	La Paz, Bolivia
DC-6B	43519	HK-1359	02/04/76	LAC Columbia	Santa Maria, Columbia
DC-6B	43520	N77DG	02/17/75	Pacific Alaska Airlines	Fairbanks, Alaska
DC-6B	43526	HP-539	04/10/72	Inair Panama	Manaus, Brazil
DC-6B	43531	TG-ADA	06/08/78	Aviateca	Guatemala City, Guatemala
DC-6B	43538	N37559	11/01/55	United Air Lines	Longmount, Colorado
DC-6B	43556	PH-DFO	08/23/54	KLM-Royal Dutch Airlines	into North Sea off The Netherlands
DC-6B	43559	SE-BDY	09/18/61	United Nations	Ndola, Zambia
C-118A	43565	HK-1705	04/29/78	LAC Columbia	Bogota, Columbia
R6D-1	43685	131582	02/25/60	U.S. Navy	mid-air with DC-3 PP-AXD, Rio de Janeiro
C-118A	43717	C-GBYA	06/26/89	Conifair Aviation	location unknown
DC-6B	43742	N8225H	01/06/60	National Airlines	Wilmington, North Carolina
DC-6B	43823	N91303	04/20/53	Western Air Lines	Oakland, California
DC-6B	43829	OO-ABG	02/18/66	Belgian Int'l Air Services	Milan, Italy
DC-6B	43834	F-ABAE	04/22/85	Securite Civile	Perpignan, France
DC-6B	43835	F-BGOD	02/20/56	Transports Aeriens Intercontintaux	Cairo, Egypt
DC-6B	43843	CF-CUP	08/29/56	Canadian Pacific Airlines	Cold Bay Alaska
DC-6B	43844	CF-CUQ	07/08/65	Canadian Pacific Airlines	Dog Creek, British Columbia
DC-6B	44057	N371	12/02/89	Gomes & Warra Aircraft Corp.	Bahamas
DC-6B	44058	N90773	09/10/61	President Airlines	Shannon, Eire
DC-6A	44069	TF-OAE	05/06/74	Fragtflug,	Nuremberg, Germany
DC-6B	44113	N6113C	11/16/70	Transportes Aereos Latin America Antigua	
DC-6B	44168	HK-1706	02/19/82	LAC Columbia	Cerro del Tablazo, Subachoque, Columbia
DC-6B	44418	I-LINE	12/18/54	Linee Aeree Italiane	Idlewild, New York
DC-6A	44421	N844TA	11/28/80	Miami Air Lease	Bimini Island, Bahamas
DC-6B	44425	N9018N	09/15/80	JMG Inc.	into sea off Haiti
DC-6B	44426	N5026K	06/22/59	Pan Am	Shannon, Ireland
C-118A	44615	N33VX	07/21/88	TACA Int'l Airlines	Golden Meadows, Louisiana
C-118A	44630	N3493F	05/05/78	Surinam Airways	Paramaribo, Surinam
C-118A	44657	N43865	07/20/79	Kimex Inc.	Kingston, Jamaica
C-118A	44659	N96040	12/08/76	Aerovias Las Minas	down between Bogota and Trinidad
C-118A	44660	N2878F	07/06/85	Northern Pacific Transport	Egelik, Alaska
C-118A	44670	HK-1702	02/10/91	Air Colombia	Bogota, Colombia
DC-6A	44678	N34954	02/01/57	Northeast Airlines	La Guardia, New York
DC-6B	44687	HK-1707X	12/08/78	LAC Colombia	Sierra Cucuy, Colombia
DC-6B	44698	CF-PWA	08/02/74	Conair Aviation	Kamloops, British Columbia, Canada
DC-6A	44914	N90779	09/18/60	World Airways	Agana, Guam
DC-6B	45064	N614SE	10/26/73	Span East Airlines	Biscayne Bay, Miami, Florida
DC-6B	45075	I-LEAD	11/24/56	Linee Aeree Italiane	Paris, France
DC-6B	45108	TR-LXN	02/26/79	Air Gabon	Monda, Gabon
DC-6B	45200	N575	08/28/58	Northwest Orient	Minneapolis-St. Paul, Minnesota
DC-6A	45243	N6118C	07/21/61	Slick Airways	Shemya, Alaska
DC-6B	45323	YS-35C	05/02/76	TACA Int'l Airways	Altaverapaz, Guatemala
DC-6B	45324	YN-BFO	12/21/87	Aeronica	Costa Rica
DC-6B	45327	N122A	08/27/78	New World Air Charter	Seeb, Oman

Type	MSN	Reg.	Date	Airline	Nearest major city
DC-6A	45369	N6541C	04/23/65	Aaxico Airlines	Mount Rainer, Washington
DC-6A	45475	N640NA	01/31/67	Saturn Airways	San Antonio, Texas
DC-6A	45476	HR-TNO	01/27/73	TAN Airlines	Tegucigalpa, Honduras
DC-6B	45479	F-BIAO	05/03/63	Air Afrique	Buca, Cameroon
DC-6A	45480	N6579C	05/10/65	Aaxico Airlines	Whiteman AFB, Missouri
DC-6A	45504	OD-AEL	03/10/66	Trans Mediterranean Airways	Athens, Greece
DC-6B	45513	CC-CCG	02/06/65	LAN Chile	Oziebrada Lo Valdes, Chile
DC-6A/C	45532	HB-IBT	05/07/69	Int'l Red Cross	Uli, Biafra
DC-6B	45537	YS-05C	06/29/88	Aerolineas El Salvador	Comalapa, El Salvador
DC-6B	45540	SX-DAE	12/08/69	Olympic Airways	Athens, Greece
Douglas DC-7					
DC-7	44275	N68N	04/00/64	Federal Aviation Admin.	Deer Valley, Arizona
DC-7	44287	N4SW	09/05/79	Butler Aircraft Co.	Klamath Falls, Oregon
DC-7	44288	N6324C	06/30/56	United Airlines	mid-air with Lockheed 1049 N6902C, Grand Canyon, Arizona
DC-7C	44872	PP-PDO	11/01/61	Panair do Brasil	Recife, Brazil
DC-7C	44876	N244B	09/05/78	Advance Aviation	Farmerville, Louisiana
DC-7C	44876	N73675	00/00/77	Consolidated Air	South Carolina
DC-7C	45071	N5903	10/09/86	T&G Aviation	Dakar, Senegal
DC-7C	45072	N5904	03/25/58	Braniff	Miami, Florida
DC-7B	45084	N815D	11/30/62	Eastern Air Lines	New York, New York
DC-7B	45085	N816D	11/28/80	Central Air Services	Pecos, Texas
DC-7C	45090	N7466	09/27/68	Universal Airlines	Cherry Point, North Carolina
DC-7C	45128	N2282	09/12/66	Airlift Int'l	Tachikawa AFB, Japan
DC-7	45142	N6328C	04/21/58	United Airlines	mid-air with F-100F s/n 56-2755, Arden, Nevada
DC-7C	45157	OO-SFA	05/18/58	Sabena	Casablanca, Morocco
DC-7C	45160	G-ARUD	03/04/62	Caledonian Airways	Douala, Cameroon
DC-7B	45192	N8210H	01/31/57	Douglas Aircraft Co.	mid-air with F-89 s/n 52-1870, Sunland, California
DC-7C	45204	N285	10/22/62	Northwest Orient Airlines	Borka Island, Alaska
DC-7C	45209	N290	06/03/63	Northwest Orient Airlines	Annette Islands, Alaska
DC-7C	45214	N8219H	00/00/80	Aero Services	Colombia
DC-7C	45229	N357AL	06/24/79	Go Transportation	Barstow, California
DC-7C	45231	I-DUVO	02/26/60	Alitalia	Shannon, Eire
DC-7B	45355	N4891C	11/15/59	Delta Air Lines	New Orleans, Louisiana
DC-7	45359	N6314J	09/11/77	Safe Air Cargo	Yakutat, Alaska
DC-7C	45366	F-BIAP	09/24/59	Transports Aeriens Intercont.	Bordeau, France
DC-7B	45452	N846D	03/10/58	Douglas Aircraft Co.	unknown
DC-7B	45455	N849D	02/08/65	Eastern Airlines	Jones Beach, New York
DC-7C	45462	N292	07/14/60	Northwest Orient Airlines	Luzon, Philippines
DC-7C	45466	N296	06/21/73	Transair	Miami, Florida
DC-7C	45467	TZ-ARC	10/04/76	Emirates Air Transport	Nairobi, Kenya
Handley Page Hermes					
Hermes I	HP.68/1	G-AGSS	12/03/45	Handley Page Ltd.	Radlett, Herts, England
Hermes IV	HP.81/7	G-ALDF	08/25/52	Airwork Ltd.	Trapani, Sicily
Hermes IV	HP.81/15	G-ALDN	05/26/52	BOAC	Atar, Libya
Hermes IV	HP.81/22	G-ALDV	04/01/58	Skyways Ltd.	Bishops Stortford, Herts, England
Lockheed 049 Constellation					
049	1968	4X-AKC	07/27/55	El Al	Petrich, Bulgarian/Greek border
C-69/049	1976	N2737A	11/08/61	Imperial Airlines	Richmond, Virginia
C-69/049	1978	N2735A	05/12/59	Capital Airlines	Charleston, West Virginia
049	2025	N86504	03/01/64	Paradise Airlines	Lake Tahoe, California
049	2026	N86505	12/28/46	Transcontinental & Western Air	Shannon, Eire
049	2028	N86507	11/18/47	Transcontinental & Western Air	Newcastle, Delaware
049	2029	N86508	05/11/47	Transcontinental & Western Air	Cape May, New Jersey
049	2032	PP-PDJ	06/12/55	Panair do Brasil	Asuncion, Paraguay
049	2034	N86510	03/29/46	Transcontinental & Western Air	Washington, D.C.
049	2035	N86511	10/01/61	Trans World Airways	Chicago-Midway Airport, Illinois
049	2037	PP-PDG	05/29/72	Amazonense Imortacao e Exportacao Ltd. Cruziero do Sul Acre, Brazil

Type	MSN	Reg.	Date	Airline	Nearest major city
049	2039	N86512	10/12/46	Transcontinental & Western Air	Newcastle, Delaware
049	2040	N86513	07/11/46	Transcontinental & Western Air	Reading, Pennsylvania
049	2046	N88846	06/22/51	Pan Am	Monrovia, Liberia
049	2047	PP-PDE	12/14/62	Panair do Brasil	Parana de Eva, Manaus, Brazil
049	2058	N88858	04/15/48	Pan Am	Shannon, Ireland
049	2062	PP-PCG	07/28/50	Panair do Brasil	Porto Alegre, Brazil
049	2066	PP-PDA	06/17/53	Panair do Brasil	Sao Paulo, Brazil
049	2083	PH-TEN	10/20/48	KLM-Royal Dutch Airlines	Prestwick, Scotland

Lockheed 649/749 Constellation

Type	MSN	Reg.	Date	Airline	Nearest major city
749	2506	VT-CQP	11/03/50	Air India	Mont Blanc, Swiss/French border
649	2521	OB-R-771	04/27/66	Lansa	Tomas, Peru
749A	2527	F-BAZX	12/24/58	Air France	Vienna, Austria
649/749A	2533	N112A	12/21/55	Eastern Airlines	Jacksonville, Florida
749	2541	PH-TER	06/23/49	KLM-Royal Dutch Airlines	Bari, Italy
749	2545	F-BAZM	01/11/63	Air France	Perpignan, France
749	2546	F-BAZN	10/28/49	Air France	St. Miguel Island, Azores
749A	2554	EI-ADA	03/13/54	BOAC	Kallang Airport, Singapore
749A	2555	N1554V	06/14/60	Pacific Northern Airlines	Mount Gilbert, Alaska
749	2558	PH-TDF	07/12/49	KLM-Royal Dutch Airline	Bombay, India
749	2560	YV-C-AMA	11/27/56	Linea Aeropostal Venezolana	Caracas, Venezuela
749	2607	HI-328	10/26/81	Aerolineas Argo	St. Thomas, Virgin Islands
749	2608	ET-T-35	07/10/57	Ethiopian Airlines	Khartoum, Sudan
749A	2611	N116	04/26/62	Federal Aviation Admin.	Canton Island
749A	2615	HH-ABA	11/11/61	Air Haiti Int'l	San Juan, Puerto Rico
749A	2616	N119A	10/19/53	Eastern Airlines	New York, New York
749A	2617	N120A	08/04/69	Trans Southern Corp.	Sao Paulo, Brazil
749A	2628	F-BAZS	08/03/53	Air France	Castellorizo, Turkey
749A	2636	N6004C	08/31/50	Trans World Airlines	Wadi Natrum, Egypt
649A	2662	5N-85H	11/28/69	Air Interamerica	Algeria
749A	2664	HK-163	08/09/54	Avianca	Terceira Island, Azores
749A	2665	XA-MEV	06/02/58	Aeronaves de Mexico	Guadalajara, Mexico
749A	2666	VT-DEP	04/11/55	Air India	Great Natuna Islands, Sarawak, South China Sea
749A	2674	F-BAZZ	09/01/53	Air France	Mont Cemet, French Alps

Lockheed 1049 Super Constellation

Type	MSN	Reg.	Date	Airline	Nearest major city
1049	4015	N6901C	03/06/66	Lineas Aereas Patagonicas Argentinas	Callao, Peru
1049	4016	N6902C	06/30/56	Trans World Airlines	mid-air with DC-7 N6324C
1049	4021	N6907C	12/16/60	Trans World Airlines	mid-air with DC-8 N8013U
1049D	4165	N6503C	11/10/58	Seaboard & Western Airlines	New York, New York
C-121C	4192	HI-515CT	04/05/90	Aerolineas Mundo	Puerto Rico
1049C	4504	PH-LKT	07/15/57	KLM-Royal Dutch Airlines	Biak, New Guinea
1049C	4509	PH-LKY	09/05/54	KLM-Royal Dutch Airline	Shannon, Eire
1049E	4512	F-BGNC	08/09/69	Catair	Douala, Cameroon
1049C	4526	N6218C	12/04/65	Eastern Airlines	mid-air with Boeing 707 N748TW, North Salem, New York
1049G	4541	N189S	05/28/63	Standard Airways	Manhattan, Kansas
1049G	4550	EC-AIN	05/05/65	Iberia	Tenerife, Canary Islands
1049G	4552	EC-AIP	03/06/61	Iberia	Sao Paulo, Brazil
1049E	4556	HK-177	01/21/60	Avianca	Montego Bay, Jamaica
1049E	4561	YV-C-AMS	06/20/56	Linea Aeropostal Venezolana	New York, New York
1049E	4564	CF-TGG	12/17/54	Trans Canada Airlines	Toronto, Canada
1049G	4575	YV-C-ANC	10/14/58	Linea Aeropostal Venezolana	Maracaibo, Venezuela
1049G	4602	D-ALAK	01/11/59	Lufthansa	Rio de Janerio, Brazil
1049G	4606	VH-EAC	08/24/60	Qantas	Plaisance Airport, Mauritius
1049G	4610	PP-VDA	08/16/57	Varig	Cabarete, Dominican Republic
1049G	4622	F-BHBC	08/29/60	Air France	Dakar, Senegal
1049H	4636	N174W	05/05/70	North Slope Supply Co.	Barrow, Alaska
1049G	4645	5T-TAC	01/00/68	Biafran Government	Port Harcourt, Biafra
1049G	4652	N7125C	11/08/60	Iberia	Barcelona, Spain
1049G	4667	VT-DIN	07/19/59	Air India	Bombay, India
1049G	4670	F-BHMK	12/06/57	Air France	Paris, France

Type	MSN	Reg.	Date	Airline	Nearest major city
1049G	4673	N8021	06/07/70	Aviaco	La Rioja, Argentina
1049H	4674	N173W	06/09/73	Aircraft Specialties Co.	Montreal, Canada
1049G	4678	HP-467	03/30/68	RAPSA	Panama City, Panama
1049H	4810	N6913C	12/14/62	Flying Tiger Line	Burbank, California
1049H	4811	N6914C	12/15/65	Flying Tiger Line	Alamosa, New Mexico
1049H	4812	N6915C	12/24/64	Flying Tiger Line	San Francisco, California
1049H	4815	N6917C	12/16/73	Andes	Miami, Florida
1049H	4817	N6921C	03/15/62	Flying Tiger Line	Guam
1049H	4822	N6920C	09/09/58	Flying Tiger Line	Mount Oyama, Japan
1049H	4824	N102R	11/23/59	Trans World Airlines	Chicago-Midway, Illinois
1049H	4827	N6923C	09/23/62	Flying Tiger Line	near Shannon, Eire
1049H	4828	CF-NAJ	08/03/69	Canairelief Air Ltd.	Uli Airstrip, Biafra
1049H	4834	N564E	10/20/71	Balair Inc.	near Great Inagua, Bahamas
1049H	4838	N566E	09/24/73	The Holy Nation of Islam,	Gary, Indiana
1049H	4840	N45516	05/11/75	Aircraft Specialties	Mesa, Arizona
1049H	4841	PH-LKM	08/14/58	KLM-Royal Dutch Airlines	near Shannon, Eire
1049H	4849	N6936C	06/22/67	Airlift Int'l	mid-air with RF-4C, Saigon, Vietnam
1049H	4850	N74CA	06/22/80	Air Traders Int'l	Columbus, Ohio
1049H	4851	N9740Z	02/03/63	Slick Airways	San Francisco, California

Lockheed 1649 Starliner

Type	MSN	Reg.	Date	Airline	Nearest major city
1649A	1002	N7301C	12/18/66	Aerocondor Colombia	Bogota, Columbia
1649A	1008	LV-GLI	09/03/64	Federal Aviation Admin.	Deer Valley, Arizona
1649A	1015	N7313C	06/26/59	Trans World Airlines	Milan, Italy
1549A	1027	F-BHBM	05/10/61	Air France	Edjele, Libya

Martin 202, 404

Type	MSN	Reg.	Date	Airline	Nearest major city
202	9134	N93050	03/07/50	Northwest Orient Airlines	Minneapolis-St. Paul, Minnesota
202	9144	N93054	01/16/51	Northwest Orient Airlines	Reardon, Washington
202	9158	N93037	10/13/50	Northwest Orient Airlines	Almelund, Minnesota
202	9159	N174A	12/01/51	Allegheny Airlines	Williamsport, Pennsylvania
202	9160	N93039	11/05/51	Transocean Airlines	Tucumcari, New Mexico
202	9161	N93040	11/07/50	Northwest Orient Airlines	Butte, Montana
202	9164	N94043	04/09/52	Transocean Airlines	O'Shuma Island, Japan
202	9165	N93044	08/29/48	Northwest Orient Airlines	Winona, Minnesota
202A	14081	N93211	01/12/55	Trans World Airlines	mid-air with DC-3 N999B, Covington, Kentucky
404	14101	N40401	08/22/62	Piedmont Airlines	Wilmington, North Carolina
404	14103	N40403	04/01/56	Trans World Airlines	Pittsburgh, Pennsylvania
404	14114	CP-1440	12/12/79	Camba Ltd.	Apolo, Bolivia
404	14116	N40412	05/30/70	Le High Acres Dev. Co.	Atlanta, Georgia
404	14120	N40416	02/19/55	Trans World Airlines	Albuquerque, New Mexico
404	14122	N445A	02/17/56	Eastern Airlines	Ownesboro, Kentucky
404	14140	N449A	07/02/63	Mohawk Airlines	Rochester, New York
404	14151	N464M	10/02/70	Golden Eagle Aviation	Silver Plume, Colorado
404	14170	N40406	11/20/66	Piedmont Airlines	New Bern, North Carolina
404	14173	N40438	11/16/79	Nevada Airlines	Grand Canyon, Arizona
404	14228	N40443	06/27/86	Frontier Airways	Buffalo, New York
404	14290	N13415	11/10/78	J. Whaley	Punta Fija, Venezuela

SAAB Scandia

Type	MSN	Reg.	Date	Airline	Nearest major city
90A-2	90.103	PP-SQE	12/30/58	VASP	Santos Dumont, Brazil
90A-1	90.106	PP-SQV	09/23/59	VASP	Sao Paulo, Brazil
90A-1	90.107	PP-SRA	11/26/62	VASP	mid-air with Cessna 310, Paraibuna, Brazil

Scottish Aviation Twin Pioneer

Type	MSN	Reg.	Date	Airline	Nearest major city
1	503	G-AOEO	12/07/57	Scottish Aviation	Tripoli, Libya
1	509	JZ-PPX	08/30/57	Nederlands Nieuw Guinea	Japen Island, New Guinea
1	511	PK-GTB	09/20/63	Garuda Indonesian Airlines	location unknown
1	517	VR-OAE	09/16/63	Borneo Airways	Siba, Sarawak
1	519	9M-ANC	05/17/67	Malaysian Airways	Limbang, Malaysia
1	523	5N-ABQ	04/04/67	Bristow Helicopters	Ughelli, Nigeria
1	526	5N-ABQ	01/16/63	Bristow Helicopters	Chepbeica, Morocco

Type	MSN	Reg.	Date	Airline	Nearest major city
Vickers Viking					
621/1A	127	G-AIJE	09/02/58	Independant Air Travel	London-Heathrow, England
614/1	142	G-AHPI	02/17/52	Hunting Air Travel	Monte la Cinta, Sicily
610/1B	148	G-AHPK	01/06/48	British European Airways	Finebush Lane, Ruislip, England
610/3B	152	G-AHPM	08/09/61	Cunard Eagle Airways	Holthei, Norway
610/1B	155	G-APHN	10/31/50	British European Airways	London-Heathrow, England
616/1B	159	VP-YEX	03/17/55	Central African Airways	Belvedere, Salisbury, England
628/1B	196	SU-AGN	03/07/58	Misrair	Port Said, Egypt
628/1B	197	OD-DLI	12/29/47	DDL	Copenhagen, Denmark
628/1B	199	OD-DLU	02/08/49	DDL	Landskrona, Denmark
632/1B	205	VT-CLY	02/06/48	Air India	Bombay, India
634/1B	210	SU-AFM	08/24/60	Don Everall Aviation	Heraklion, Crete
634/1B	213	SU-AFK	12/15/53	Misrair	Cairo, Egypt
610/1B	216	F-BJER	09/11/63	Airnautic	Pyrenees, France
610/1B	229	C-AIVP	04/05/48	British European Airways	mid-air with Yak, Berlin, Germany
644/1B	231	YI-ABQ	10/10/55	Iraqi Airways	Baghdad, Iraq
610/1B	241	G-AJBO	05/01/57	Eagle Aviation Ltd.	Blackbushe, Hampshire, England
610/1B	248	F-BFDN	09/05/59	Airnautic	Traliceto, Southern Corsica
610/1B	262	G-AJDL	01/05/53	British European Airways	Nutts Corner, Belfast, Northern Ireland

APPENDIX III

Civil DC-3 Accidents And Incidents In The Continental United States

The wreckage of United DC-3A N16073, msn 1913, is hauled away after crashing into the San Francisco Bay on Feb. 9, 1937. The crash was caused by the accidental jamming of the elevator controls when a microphone was dropped in the cockpit. The microphone fell under the yoke and jammed the controls in the nose down position during landing. There was not enough altitude for the crew to dislodge the microphone and recover the aircraft's attitude.

Reg.	Msn	Year	Location	Airline		Date	Rebuilt
N16073	1913	1937	San Francisco CA	United	USA	02/09	
N16070	1910	1937	Cheyenne WY	United	USA	04/27	Yes
N16074	1914	1937	Knight WY	United	USA	10/17	
N18108	1956	1938	Cleveland OH	United	USA	05/24	
N17316	1931	1938	Albuquerque NM	TWA	USA	08/24	Yes
N16066	1906	1938	Point Reyes CA	United	USA	11/29	
N18105	1953	1940	Moline IL	United	USA	01/12	Yes
N21789	2188	1940	Lovettsville VA	Penn Central	USA	08/31	
N16086	1925	1940	Centerville UT	United.	USA	11/04	
N25678	2175	1940	Chicago IL	United	USA	12/04	
N16015	1553	1940	St Louis MO	American	USA	12/11	Yes
N17315	1930	1941	St Louis MO	TWA	USA	01/23	
N28394	3250	1941	Atlanta GA	Eastern	USA	02/26	
N21727	2143	1941	Vero Beach FL	Eastern	USA	04/03	
N16064	1904	1941	Denver, CO	United	USA	06/10	Yes
N1948	3298	1941	Harrisburg PA	TWA	USA	06/16	Yes
N25691	2256	1941	Morgantown WV	Penn Central	USA	10/02	Yes
N21712	2124	1941	Moorehead MN	Northwest	USA	10/30	
N1946	3295	1942	Las Vegas NV	TWA	USA	01/16	
N17322	1918	1942	Little Rock AR	American	USA	02/05	Yes
N18143	2139	1942	La Guardia NY	American	USA	02/17	Yes
N21788	2187	1942	Detroit MI	Penn Central	USA	03/02	
N16064	1904	1942	La Guardia NY	United	USA	04/11	Yes
N18146	1978	1942	Salt Lake UT	United	USA	05/01	
N21714	2129	1942	Miles City MT	Northwest	USA	05/12	
N16018	1556	1942	Little Rock, AR	American	USA	05/21	Yes
N17320	1966	1942	Dayton OH	TWA	USA	06/27	Yes
N21711	2123	1942	Bismark ND	Northwest	USA	08/29	Yes
N16017	1555	1942	Palm Springs CA	American	USA	10/23	

W.T. Larkins Collection

Reg.	Msn	Year	Location	Airline		Date	Rebuilt
N18951	2015	1942	Kansas City KS	TWA	USA	11/04	
N16064	1904	1942	Dayton OH	United	USA	11/18	
N16060	1900	1942	Fairfield UT	Western	USA	12/15	
N16090	1929	1943	Oakland CA	United	USA	01/11	
N16014	1552	1943	Trammel KY	American	USA	07/28	
N33657	4803	1943	Dallas TX	American	USA	09/15	
N16008	1588	1943	Centerville TN	American	USA	10/15	
N19968	3252	1943	New Orleans LA	Eastern	USA	11/19	
N21767	2166	1944	Memphis TN	American	USA	02/10	
N28310	2251	1944	Hanford CA	TWA	USA	11/04	
N17322	1968	1944	Van Nuys CA	TWA	USA	12/01	
N21752	2165	1944	Saline MI	American	USA	12/24	Yes
N25684	2215	1945	Burbank CA	American	USA	01/10	
N18142	2138	1945	Rural Retreat VA	American	USA	02/23	
N25692	2262	1945	Morgantown WV	Penn Central	USA	04/14	
`N21914	4965	1945	Lynchburg VA	American	USA	04/28	Yes
N25647	2235	1945	Florence SC	Eastern	USA	07/12	
N33631	4137	1945	Florence SC	Eastern	USA	09/07	
N19939	4992	1945	Western Springs IL	TWA	USA	09/29	Yes
N18123	1999	1945	La Guardia NY	Eastern	USA	12/30	
N21786	4131	1946	Birmingham AL	Penn Central	USA	01/06	
N19970	3254	1946	Cheshire CT	Eastern	USA	01/18	
N25675	2147	1946	Elk Mt WY	United	USA	01/31	
N14941	4468	1946	Newark NJ	Air Cargo Tsp	USA	02/25	Yes
N21799	2203	1946	El Centro CA	American	USA	03/03	
N33621	3286	1946	Hollywood CA	Western	USA	04/24	
N53218	19626	1946	Richmond VA	Viking	USA	05/16	
N28383	4091	1946	Chicago IL	TWA	USA	07/02	
N50040	19980	1946	Lumberton NC	Trans Carib	USA	08/08	
N51878	4542	1946	Moline IL	Trans Luxury	USA	08/21	
N88826	19776	1946	Ashland MS	American	USA	08/25	
N57850	9214	1946	Elko NV	Trans Luxury	USA	09/05	
N38942	12971	1946	Laramie WY	Nat Air Transport	USA	10/17	
N20750	4993	1946	Meridian MS	-----	USA	11/10	
N19947	4873	1946	Cleveland OH	United	USA	11/11	
N18645	11662	1946	Lebec CA	Western	USA	11/13	
N45395	11642	1946	Mt Laguna CA	Western	USA	12/24	
N15577	4805	1946	Michigan City IN	American	USA	12/28	
N88873	13640	1946	Charleston SC	Intercontinental	USA	12/31	
N21746	2104	1947	Jones Beach NY	American	USA	01/05	Yes
N50046	34373	1947	Vineland NJ	Nationwide Air	USA	01/05	
N88872	3274	1947	Galax VA	Eastern	USA	01/12	
N54550	9965	1947	League City TX	Atlantic & Pac	USA	02/14	
N33646	4125	1947	Philadelphia PA	United	USA	02/24	
N88804	9921	1947	Charleston SC	USAir	USA	03/14	
N52710	19095	1947	Lantana FL	Int Air Frt	USA	04/04.	
N49657	9066	1947	Columbus GA	Delta	USA	04/22.	
N53196	19435	1947	Newark NJ	Union Southern	USA	05/04	
N19917	7320	1947	Hartfield, VA	Capital	USA	05/08	Yes
N79024	9887	1947	Melbourne FL	Burke Air	USA	07/13	
N88787	19639	1947	La Guardia NY	American	USA	08/08	
N95486	6065	1947	Yakutat AK	Columbia	USA	11/27	
N60331	12968	1948	Savannah River GA	Coastal	USA	01/07	
N28384	4092	1948	Oxon Hill MD	Eastern	USA	01/13	
N206	4776	1948	Navajo Peak CO	CAA	USA	01/21	
N36480	33598	1948	Coalinga CA	Airline Trnspt	USA	01/28	
N36498	12527	1948	Columbus OH	Bruning Avn	USA	02/25	
N64722	10052	1948	San Jose CA	Eagle Air Frt	USA	03/08	
N79042	9394	1948	Elroy AZ	Eagle Air Frt	USA	06/12	
N17335	1921	1948	Burlington VT	Colonial	USA	09/20	
N66637	11800	1948	Cape Spencer AK	Pacific Alaska	USA	11/04	
N79025	10181	1949	Seattle WA	Seattle Air	USA	01/02	
N53210	13777	1949	Brandywine IN	Coastal Cargo	USA	01/06	
N91006	20208	1949	Homer AK	Alaska	USA	01/20	

Reg.	Msn	Year	Location	Airline		Date	Rebuilt
N17713	4582	1949	Oakland CA	United	USA	01/27	
N16002	1496	1949	off Miami FL	Airborne Trnspt	USA	04/26	
N19963	2260	1949	Chesterfield NJ	Eastern	USA	07/30	
N53594	20193	1949	Teterboro NJ	Meteor Air Trnspt	USA	10/29	Yes
N29086	4304	1949	Akron OH	Harrington's Inc	USA	11/04	
N54337	20136	1949	Detroit MI	Meteor Air Trnspt	USA	11/19	Yes
N60256	9201	1949	Benicia CA	Arrow	USA	12/07	
N25691	2256	1949	Washington DC	Capital	USA	12/12	
N18936	2011	1950	Utica NY	Robinson	USA	09/04	
N19928	7400	1951	Sioux City IA	Mid Cont	USA	03/02	
N63439	20229	1951	Santa Barbara CA	Southwest	USA	04/06	
N16088	1927	1951	Ft. Wayne IN	United	USA	04/28	
N25646	2234	1951	Ocala. FL	Eastern	USA	11/27	Yes
N17109	4999	1951	Denver CO	United	USA	12/04	
N41748	11827	1952	Mt Crillon AK	General	USA	01/12	
N65384	18984	1952	Greenboro NC	Piedmont	USA	10/20	Yes
N21716	2131	1952	Richmond IN	Lake Central	USA	12/15	Yes
N41455	11876	1953	Miles City MT	General	USA	02/15	Yes
N65743	20432	1953	Selleck WA	Miami	USA	04/14	Yes
N28345	2224	1953	Marshall TX	Delta	USA	05/17	
N53596	22433	1953	St Louis MO	Meteor Air Trnspt	USA	05/24	Yes
N15569	4887	1953	Los Angeles CA	Western	USA	06/29	
N19941	6333	1953	Vail WA	Regina	USA	09/01	
N49551	4940	1954	Kansas City MO	Zantop	USA	01/20	
N43V	42958	1954	Bristol, TN	Piedmont	USA	02/26	Yes
N51359	13759	1954	Atlanta GA	Delta	USA	06/15	
N91008	25422	1954	MacGrath, AK	Alaska	USA	08/09	
N61451	4630	1954	Mason City IA	Braniff	USA	08/22	
N17891	11745	1954	Mt Success	Northeast	USA	11/30	
N24320	20197	1954	Pittsburgh PA	Johnson Flt Svc	USA	12/22	Yes
N999B	4255	1955	Covington, KY	Castletown	USA	01/12	
N19942	4872	1955	Lebanon, NH	Northeast	USA	05/31	Yes
N51167	7384	1955	Kansas City KA	TWA	USA	07/12	Yes
N45333	6330	1955	Yakima WA	Northwest	USA	07/15	Yes
N18945	2018	1955	Hobbs NM	Continental	USA	08/29	Yes
N74663	6257	1955	Burbank CA	Curry Air Trans.	USA	09/08	
N62374	12534	1956	Pullman WA	W. Coast Airlines	USA	02/26	
N45V	18984	1956	Shelby NC	Piedmont	USA	06/13	Yes
N33315	4978	1956	Bartlesville OK	Continental	USA	09/09	Yes
N65276	19202	1957	Phoenix AZ	Frontier	USA	04/21	Yes
N88835	19448	1957	Clarksburg MD	Capital	USA	06/22	
N34417	7337	1957	New Bedford MA	Northeast	USA	09/15	
N3947A	9856	1957	Charlottsville VA.	Piedmont	USA	10/02	Yes
N49553	4820	1958	Martinsburg WVA	Capital	USA	06/04	
N64424	4424	1958	Pueblo CO	Frontier	USA	08/23	Yes
N18941	2007	1959	Tri Cities TN	Southeast	USA	01/08	
N17314	1924	1959	Kerrville TX	General	USA	02/01	
N44993	6260	1959	Charleston WV	Capital	USA	08/26	Yes
N67589	19656	1959	Santa Maria CA	Pacific Airlines	USA	10/26	
N55V	20477	1959	Waynesboro VA	Piedmont	USA	10/30	
N38G	4759	1959	Cleveland OH	National Aero	USA	11/05	
N200	20400	1961	Mustang OK	FAA	USA	04/21	
N58731	13532	1962	Big Trees CA	Blatz	USA	01/20	Yes
N3588	20178	1962	Dallas TX	Perdue Aeron	USA	04/18	
N386T	20411	1963	Morgantown WV	Purdue Aeron	USA	11/29	
N65SA	4925	1964	Huntsville AL	Southern	USA	2/12	Yes
N401D	4970	1964.	Chicago IL	Snow Valley	USA	03/09	
N61442	9642	1964	Miles City MT	Frontier	USA	03/12	
N91003	9708	1964	Pueblo CO	Central	USA	07/20	
N61350	4535	1964	Tonopah NV	Hawthorne NV	USA	08/19	
N91016	11853	1965	Nikolski AK	Reeve	USA	05/29	
N123H	12679	1965	Stockton CA	Biffs	USA	07/02	Yes
N485	4848	1965	Salt Lake UT	Edde	USA	11/27	
N17337	1962	1966	Long Beach MS	Aeronaut	USA	06/27	

Reg.	Msn	Year	Location	Airline		Date	Rebuilt
N4994E.	12442	1966	Gallup NM	Frontier	USA	07/27	
N54370	19220	1967	Merced CA	Turkey Farms	USA	03/24	
N65276	19202	1967	Denver CO	Frontier	USA	12/21	
N6898D	20082	1968	Vichy MO	Purdue	USA	06/28	Yes
N15570	6320	1969	Hawthorne NV	Mineral County	USA	02/18	
N142D	1946	1969	N Orleans LA	Avion Inc	USA	03/20	
N154R	6156	1970	McGrath AK	Reeder	USA	07/03	
N47	33155	1970	Anchorage AK	FAA	USA	10/01	
N14273	7366	1971	Shawnee OK	Patterson Lsg	USA	05/14	
N90627	4642	1971	Shelter Cove CA	Greyhound Lsg	USA	06/27	
N44992	11738	1971	Peotone, IL	Unknown	USA	09/06	Yes
N15HC	43080	1974	Houston, TX	Horizon Prop	USA	02/01	
N9BC	9510	1975	Ft Lauderdale FL	Air O'Hare	USA	01/01	
N6	4146	1975	Jefferson PA	FAA	USA	03/27	
N144A	9723	1975	Wakeman, OH	Stoney's	USA	09/11	
N73KW	2252	1977	Miami FL	Air Sunshine	USA	01/15	
N38AP	10217	1977	Newton GA	Hall Mining	USA	06/13	
N51071	4837	1977	Evansville IN	Air Indiana.	USA	12/13	
N45873	12458	1978	Richmond IN	Air Sales	USA	07/09	
N74Z	25953	1978	Ft Walton Bch FL	Evergreen	USA	10/01	
N41447	12987	1978	Des Moines IA	SMB Stage	USA	12/02	
N25656	4845	1978	Lorida FL	Caribe Air	USA	12/06	Yes
N133AC	6260	1978	Port Mayaca FL	Acft Sales	USA	12/11	
N4996E	12141	1978	Belle Glade FL	Rodriguez	USA	12/14	
N9025R	4998	1979	South Bay FL	Waggoner	USA	01/10	
N148Z	20422	1979	Selway River ID	USFS	USA	06/11	
N427W	43073	1979	Ft. Lauderdale, FL	Bradley Avn	USA	06/12	
N21782	2170	1979	Belle Glade FL	M J Supply	USA	07/13	
N59314.	12363	1979	Bettles AK	J L Dodson	USA	07/27	Yes
N99663.	33467	1979	Bettles AK	K.B. Martin	USA	10/30	
N64490	25360	1980	Athol IN	Hackney	USA	04/11	
N709Z	27182	1980	Ft Lauderdale FL	Florida Equity	USA	04/24	
N45864	9304	1980	New Smyrna FL	Rule	USA	08/01	
N3WX	34373	1980	Miami, FL	Air Miami	USA	12/22	Yes
N87652.	25534	1981	Miami FL	S.S Airways	USA	01/17	Yes
N3433P	4346	1981	Laredo TX	Atkins Avn	USA	04/01	Yes
N111ST	4661	1981	Pilot Point AK	Acft Svces	USA	06/01	
N18949	2013	1981	Melbourne FL	Hawthorne Coll	USA	06/24	
N129H˘	4126	1982	Sandwich, IL	M Borghorst	USA	05/29	Yes
N95C	20139	1982	St Petersburg FL	Fromhagen	USA	06/06	
N38941	6332	1983	Springfield, MO	Crystal Shamrock	USA	03/24	
N95460	20190	1983	Fort Yukon AK	Yukon Air Serv	USA	06/24	Yes
N70003	12938	1984.	Memphis TN	River City	USA	08/11	
N139PB	2239	1984	Naples, FL	P.B.A.	USA	12/27	Yes
N28BA	4583	1985	Charlotte NC	S J Bowman	USA	02/05	Yes
N157U	4132	1985	Perris CA	B Conatser	USA	05/04	
N168Z	20850	1985.	King Salmon AK	Northern Pen	USA	06/30	
N777YA	25634.	1985	Dillingham, AK	L.L. Bingman	USA	12/15	Yes
N711Y	13658	1985	De Kalb TX	Century Equity	USA	12/31	
N25CE	12476	1986	El Paso TX	Atorie Air	USA	04/05	Yes
N3433U	43087	1986	Las Vegas NV	Atorie Air	USA	05/30	Yes
N31MS	11631	1987	Hilliard FL	F.J. Bailey	USA	06/30	Yes
N39DT	4871	1987	Laredo TX	Ferreteria	USA	07/28	
N28889	20520	1989	Summerland FL	Monroe Cty	USA	05/02	
N47E	13456	1989	Waterman IL	Condor Ent	USA	05/22	
N8042X	19041	1989	Scow Bay AK	CA Air Tours	USA	06/26	Yes
N143JR	1995	1989	Columbus IN	Rhoades Int'l	USA	09/15	Yes
N1FN	11685	1990	Caron Bridge WV	K & K Acft	USA	05/19	
N13JA	4082	1991	Apple Creek OH	JAARS	USA	08/24	Yes
N8056	26008	1993	Zepherhills FL	Phoenix Air	USA	04/20	
N3433Y	43089	1994	Spokane WA	Salair	USA	03/18	
N54NA	19475	1995	Whitesville NY	Business Air	USA	07/19	

APPENDIX IV
Obtaining A USGS Topo Map

The United States Geological Survey publishes topographical maps for every inch of the United States. The maps range in cost from a minimum of $2.50 to as high as $10 and are available at many bookstores, blueprint stores, engineering supply stores, main libraries, and university libraries. Consult your local phone book.

To order maps directly from the USGS, you must first obtain the following items for the state in which your wreck is located:

- *Catalog of Topographical and Other Published Maps*
- *Index to Topographic and Other Map Coverage*

Both publications are free of charge. To order the above publications, write to:

Map Distribution/Sales
U.S. Geological Survey
Box 25286, Federal Center
Denver, Colo. 80225

The ***Catalog of Topographical and Other Published Maps*** will provide you with a list of USGS map dealers in your state. Once you have selected your map, you can order by mail from the above address or visit one of the retailers listed in the map catalog. Be sure to pick up the ***Guide to Topographic Map Symbols***.

Maps must be ordered by map name, state, and series/scale. Payment should be made by check or money order made payable to the **Department of the Interior — USGS.**

APPENDIX V
Further Sources

The source names listed below should be consulted in addition to those listed in Volume I, *Wreck Chasing: A Guide To Finding Aircraft Crash Sites*.

Aireports
5606 Forest Lake Drive
San Antonio, Texas 78244
+1 (210) 661-6989

Aireports offers a full line of CAB/NTSB aircraft accident reports from the 1950s and '60s. These reports have been painstakingly retyped to compensate for the poor quality of the originals. Prices vary, so send a SASE for the current list.

Pacific Aero Press
P.O. Box 20092
Castro Valley, Calif. 94546-8092
+1(415) 593-7024, Fax (415) 595-4831

Publishers of Volume I, *Wreck Chasing: A Guide To Finding Aircraft Crash Sites*. Send a SASE for information on other items of interest to *Wreck Chasers*.

Wildflower Productions
4104 — 24th St., Suite 530
San Francisco, Calif. 94114
+1 (415) 282-9112

This company has transferred all of the USGS topo maps for the San Francisco Bay Area onto CD-ROM. Easy-to-use tools let you browse, measure, mark and print or export maps from the disk. Wildflower is on line; e-mail at: INFO@TOPO.com, or they have a home page at: HTTP://www.topo.com.

Research Systems
65 North, 3700 West
Hurricane, Utah 84737

Offers a data base with over 300 CAB and NTSB reports available. Send SASE with information request.

APPENDIX VI

This worksheet has been reduced by 70 percent to fit on this page. If it is enlarged by 30 percent it will fill an 8 1/2 inch-by-11 inch sheet of paper. Using this worksheet will help in your Wreck Chasing efforts. Refer to *Part I, Researching An Airliner Crash From Start To Finish* for more information.

Make/Model	_____ [Type]
Registration/Serial	_____ Previous Regs. _____
MSN	_____ Date of Delivery _____
Operator	❏ Civilian:_____ ❏ USN ❏ USAAF/USAF
Crash Date	___/___/___
Location	_____
7.5 Min Quad	_____
Accident Report	❏ FAA ❏ USN ❏ AAF/USAF Report#/Date_____
	date requested (___/___/___)
Aircraft History	❏ FAA ❏ USN ❏ AAF/USAF
	date requested (___/___/___)
Newspapers	❏ _____
	Date ___/___/___ Page _____
	Date ___/___/___ Page _____
	Date ___/___/___ Page _____
	Date ___/___/___ Page _____
	❏ _____
	Date ___/___/___ Page _____
	Date ___/___/___ Page _____
	Date ___/___/___ Page _____
	Date ___/___/___ Page _____
	❏ _____
	Date ___/___/___ Page _____
	Date ___/___/___ Page _____
	Date ___/___/___ Page _____
	Date ___/___/___ Page _____
Witnesses	❏ Mr/Mrs_____ ❏ Mr/Mrs_____
	❏ Mr/Mrs_____ ❏ Mr/Mrs_____

Reported Location (from newspaper descriptions)
_____
_____
_____

BIBLIOGRAPHY AND SUGGESTED READING

Davies, R.E.G. *Delta: An Airline And Its Aircraft.* 1990. Paladwr Press: McLean, Va.

Davis, John M. and Martin, Harold G. and Whittle, John A. *The Curtiss C-46 Commando.* 1978. Air Britain (Historians) Ltd: Tonbridge, Kent, England.

Derogee, Erik and Simon, Paul and van Stelle, Peter. *Airlines & Airports, Coding & Decoding.* 1994. EPS Aviation Productions: Woerden, The Netherlands.

Eastwood, A.B. and Roach, J. *Piston Engine Airliner Production List.* 1991. The Aviation Hobby Shop: West Drayton, Middlessex, England.

Eddy, Paul and Potter, Elaine and Page, Bruce. *Destination Disaster.* 1976. Quadrangle: New York, N.Y.

Edwards, Allan. *Flights Into Oblivion.* 1993. Paladwr Press: McLean, Va.

Frank, Beryl. *Plane Crashes.* 1981. Bell Publishing: New York, N.Y.

Gero, David. *Aviation Disasters: The World's Major Civil Airliner Crashes Since 1950.* 1993. Patrick Stephens Ltd: Sparkford, Nr. Yeovil, Somerset, England.

Gradidge, J.M.G. *The Douglas DC-3 And Its Predecessors.* 1984. Air Britain: Tonbridge, Kent, England.

Grayson, David. *Terror In The Skies.* 1988. Citadel Press: Secaucus, N.J.

Job, Macarthur. *Air Disaster Vol. I.* 1994. Aerospace Publications: Sydney, Australia.

Juptner, Joseph P. *U.S. Civil Aircraft,* Vol. 1 through Vol. 9. 1962. Aero Publishers: Fallbrook, Calif.

Macha, Gary Patric. *Aircraft Wrecks in the Mountains and Deserts of California.* 1991. Aircraft Archaeological Press: Huntington Beach, Calif.

McLachlan, Ian. *Final Flights.* 1989. Patrick Stephens Ltd: Great Britian.

Nance, John J. *Blind Trust.* 1986. William Morrow & Co: New York, N.Y.

Roach, J.R. and Eastwood, A.B. *Jet Airliner Production List.* 1992. The Aviation Hobby Shop: West Drayton, Middlessex, England.

Roach, John and Eastwood, A.B. *Turbo Prop Airliner Production List.* 1994. The Aviation Hobby Shop: West Drayton, Middlessex, England.

Serling, Robert. *Howard Hughes' Airline: An Informal History of TWA.* 1983. St. Martins: New York, N.Y.

Stewart, Stanley. *Air Disasters.* 1986. Ian Allen Ltd: London, England.

Stewart, Stanley. *Air Disasters.* 1986. Barnes & Noble Inc: New York, N.Y.

Stich, Rodney. *Unfriendly Skies.* 1990. Diablo Western Press: Reno, Nev.

Taylor, Michael J.H. *Janes Encyclopedia of Aviation.* 1989. Portland House: New York City, N.Y.

Veronico, Nicholas A., et al. *Wreck Chasing: A Guide To Finding Aircraft Crash Sites,* 1993. Pacific Aero Press: Castro Valley, Calif.

Watt, Ronnie. *Helderberg Death Flight SA295.* 1990. Southern Books: Johannesburg, S. Africa.

Periodicals

Airliners: The World's Airline Magazine. World Transport Press. Miami, Fla. Various issues.

ABOUT THE AUTHORS

NICHOLAS A. VERONICO

Nicholas A. Veronico is an aviation historian interested in aircraft crash sites. He has written more than 600 aviation-related newspaper and magazine articles. Veronico has authored or co-authored nine books, most recently *F4U Corsair; Wreck Chasing: A Guide To Finding Aircraft Crash Sites; Fly Past, Fly Present: A Celebration of Preserved Aviation;* and *Douglas DC-3: 60 Years and Counting*. Nicholas A. Veronico serves as the managing editor of *Airliners: The World's Airline Magazine*.

ED DAVIES

Born in Wales, Ed Davies graduated as an electrical engineer and completed two years full time national service as a sub lieutenant in the Fleet Air Arm of the Royal Navy. He worked as an engineer in a nuclear power plant before emigrating to North America in 1966. Davies held various managerial positions in the aluminum industry before branching out in his own business in 1983. He retired two years ago to devote more time to his hobby of aviation research in general, and the history of surviving DC-3s in particular. He has written many articles for the aviation press in the U.K. and the United States and has co-authored *Wreck Chasing A Guide To Finding Aircraft Crash Sites,* and *Douglas DC-3: 60 Years and Counting*.

DONALD B. MCCOMB JR.

Donald B. McComb Jr. is a lifelong aviation enthusiast and an airline accident specialist. He is a noted collector of commercial and military aircraft components ranging from airliner doors to aircraft instruments. He also has a collection of parts recovered from several airline accident sites. A native of California, he now resides in Miami, Fla.

MICHAEL B. MCCOMB

Michael B. McComb specializes in aviation archaeology of remote airline accidents sites. He founded Research Systems in 1990, an organization devoted to technical research and preservation of both commercial and military aircraft wreckage sites. McComb also holds an Airline Transport Rating and has accumulated more than 7,000 hours as a commercial pilot. He currently resides with his family in southern Utah.

Also Available, Volume I:

WRECK CHASING

A Guide To Finding Aircraft Crash SItes

- This first volume is a complete guide to finding and locating aircraft crash sites. A particular emphasis has been placed on locating military aircraft of World War II.
- Commercial aircraft crashes detailed include Pan Am's Martin M-130 "Philippine Clipper" and British Commonwealth Pacific Airlines' DC-6 "Endeavour."
- Other aircraft profiled include a Bell P-63 Kingcobra, a Bell P-39 Airacobra recovered from a Canadian lake, Vought F4U Corsair, Grumman F3F recovered from the waters off San Diego, California, Consolidated B-24 Liberator, airship USS *Macon,* McDonnell F2H Banshee, and a North American P-51H.
- Search for Pearl Harbor veterans in Hawaiian waters.
- A review chapter takes the reader from start to finish in finding a crashed aircraft.
- How to obtain U.S. Navy pilot history cards of aircraft accidents.
- Obtaining U.S. Air Force aircraft history cards.
- Latitude and longitude listing for 500 military and commercial aircraft crash sites.
- 84 pages, 64 black & white illustrations, softbound, 8 1/2 inches by 11 inches.
- Authored by Nicholas A. Veronico, Ed Davies, A. Kevin Grantham, Enrico Massagli, Thomas Wm. McGarry, and Walt Wentz.
- Published by Pacific Aero Press.

What others have said about **WRECK CHASING**: A Guide To Finding Aircraft Crash Sites:
   "...a goldmine of information and also...a fascinating document for the final resting places of many aircraft." — **Air Classics**
   "...this is a well-written 'how-to' (and often a wise 'how not to') approach to researching a wreck, finding and exploring a crash site, and a list of crash site courtesies." — **Airliners**

Available from finer booksellers, or contact:

World Transport Press
P.O. Box 521238, Miami, Fla. 33153-1238
Tel. +1(305) 477-7163 • Fax +1(305) 599-1995

$11.95
item 2231